T0407367

SEX IS A SPECTRUM

Sex Is a Spectrum

THE BIOLOGICAL LIMITS OF THE BINARY

AGUSTÍN FUENTES

PRINCETON UNIVERSITY PRESS
PRINCETON & OXFORD

Published by Princeton University Press
41 William Street, Princeton, New Jersey 08540
99 Banbury Road, Oxford OX2 6JX

press.princeton.edu

ISBN 9780691249414
ISBN (e-book) 9780691249421

British Library Cataloging-in-Publication Data is available

Editorial: Alison Kalett and Hallie Schaeffer
Production Editorial: Natalie Baan
Jacket Design: Chris Ferrante
Production: Jacquie Poirier
Publicity: Kate Farquhar-Thomson and Julia Haav
Copyeditor: Hank Southgate

This book has been composed in Arno

Printed in the United States of America

10 9 8 7 6 5 4 3 2 1

CONTENTS

ACKNOWLEDGMENTS

GOOD SCIENCE and the best scholarship are collaborative. This book is possible due to the amazing work of incredible scholars and their willingness to share insights, discoveries, methods, and challenges. I am especially indebted to Anne Fausto-Sterling and Sarah Richardson for their monumental contributions and for facilitating my and the scholarship of so many others on the topic of sex. Other scholars who are central to my learning, thinking, and working in this area include Sari Van Anders, John Archer, Samantha Archer, Bruce Bagehemil, Rick Bribeiscas, Kate Clancy, Catherine Clune-Taylor, Lucy Cooke, Amanda Cortez, Kimberly Crenshaw, Molly Crockett, Simone de Beauvoir, Andrea DiGiorgio, Phyllis Dolhinow, L. Zachary DuBois, Holly Dunsworth, Lise Eliot, Cordelia Fine, Richard Frum, Lee Gettler, Patricia Gowaty, Scott Gilbert, Matthew Gutmann, Donna Haraway, bell hooks, Sarah Hrdy, Daphna Joel, Rebecca Jordan-Young, Katrina Karkazis, Craig Kirkpatrick, Alexandra Kralick, Sarah Lacy, Roger Lancaster, Judith Lorber, Katherine C. MacKinnon, Donna Maney, Jon Marks, Jim McKenna, Stephanie Meredith, Mary Midgley, Mia Miyagi, Monique Borgerhoff Mulder, Serena Nanda, Robin Nelson, Cara Ocobock, Gina Rippon, Joan Roughgarden, Julienne Rutherford, Heather Shattuck-Heirdorn, Rick Smith, Zuleyma Tang-Martinez, Caroline VanSickle, Cara Wall-Scheffler, Polly Wiessner, Mary Jane West-Eberhard, Adrienne Zihlman, Marlene

Zuk, the GENDERSCI lab at Harvard, and many others. I thank my amazing colleagues in the Department of Anthropology at Princeton for their support and intellectual community and all the students at UC Berkeley, Central Washington University, The University of Notre Dame, and Princeton University who have inspired me to think and learn.

I effusively thank Rebecca Jordan-Young and Donna Maney for their generous, constructive, insightful, and comprehensive reviews on earlier versions of the manuscript and an anonymous reviewer for their strong critique and extensive commentary. This book would not have been possible without the amazing Alison Kalett, Hallie Schaeffer, and the entire team at PUP, and the wonderful support of Melissa Flashman at Janklow & Nesbit Associates. Hank Southgate did a terrific copyedit. My family, especially my mother and sisters, have shaped my engagement with these topics, and Shelley the Wonderdog offered support throughout the writing process. Finally, and most importantly, this book and my capacity to write it results in large part from the more than thirty years of partnership, discussions, debates, and collaborations with Devi Snively.

SEX IS A SPECTRUM

Introduction

SEX BIOLOGY IS INTERESTING

IMAGINE YOU are a fish called the bluehead wrasse, living off the coast of Florida. As you grow up, you, just like all to the other bluehead wrasse your age and size, develop one set of reproductive organs. You are what we'd call female, so you produce eggs. There is only one very large member of your group, and they are the group male, so produce sperm. But over the next few weeks you grow really fast, becoming the second-largest fish on your reef. Then the male gets eaten. Almost immediately your body starts to change, your reproductive organs mold, shift, and alter their form. *You* become the group's sperm producer. As a bluehead wrasse, you can have one body and one set of DNA, but multiple forms of reproductive biology across your lifetime.

Bluehead wrasse reproductive biology is not the most common pattern in the animal kingdom, but it's also not that weird. When most people think of the biology of reproduction, they typically envision two fixed kinds in each species: female and male. This is (mostly) right when it comes to the reproductive organs themselves, but not accurate for entire bodies and lives.

Most species do have two types of reproductive organs, and they are often found in two slightly different forms of that species' body plan. But not always. As with the wrasse, many fish start out with one set of reproductive organs, and once they grow to a certain size, they redo their anatomy and develop a new set of reproductive organs. Each earthworm's body has both types of its species' reproductive organs. Bees have two kinds of reproductive organs across three kinds of bodies. All mammal mothers lactate, but in some species of bat, fathers do too. And, as in the two different types of orangutan male, one with big face flanges and the other without, there can also be quite a bit of variation in bodies and behavior even among those individuals within the same species that have the same reproductive organs.

There is an explosion in research on the biology of reproduction—what we'll call sex biology—in the animal kingdom. While we continue to find that there are important differences in reproductive biology producing female, male, and sometimes intersex bodies in any given species, there is also a lot of variation, and overlap, in the actual biology and behavior that make up these categories. The variation we observe across the animal kingdom doesn't represent unusual exceptions to some kind of rule of sex; rather, this spectrum of variation tells us that females and males are not two different kinds of thing. Sex biology is not about two distinct kinds, a binary; instead, it's about patterns of variation in bodies, behavior, and lives that differ, overlap, and intertwine. Sex biology, as it turns out, makes life quite interesting.

The explosion in research is not limited to how other animals "do sex." There is also enormous investigation into human bodies, reproductive processes and patterns, health, hormones, genitals, genetics, behavior, and other related topics. For example, we now know that human brains don't come in "male" and "female" versions. Also, unlike some other mammals, all human caretakers (regardless of their reproductive organs) can undergo changes in their brains, bodies, and behavior when they take care of babies. Fascinatingly, human sexual behavior, including the targets of attraction and arousal, is not necessarily linked to what kind of reproductive organs one has. And, most importantly, human sex is never just about biology; we have gender too.

In short, there is a lot going on in science regarding sex and gender in humans. Unfortunately, there is also a ton of misunderstanding in society about what biology, especially sex biology, tells us and what it doesn't tell us. And there is lack of awareness of just how diverse and variable humans are. To better understand biology and sex in humans, we need to learn about our bodies, histories, cultures, and behavior. We have to understand what it means that everything about humans is a supercomplicated blend of biology and culture. We need to combine our knowledge of biology, sex, and the human experience into a new narrative. My goal in this book is to put forward this new narrative and show how the biology of sex actually works, what it does and does not tell us, and how we might incorporate this knowledge into our education, lives, and laws.

To do so, I will first summarize what is currently known about the biology of sex in animals and how, and why, that relates to humans. This is important because understanding animal biology is at the heart of understanding human biology (we are, after all, animals). From there, I will illustrate what we know about sex, in biology and behavior, in the human past and present, across the last two million years of our lineage's existence and among the eight billion humans living today. What this knowledge from the animal world, the human past, and the human present shows us is that biology as it relates to sex is not binary, meaning that it does not come in two distinct kinds: male and female. This is not to say that females and males are the same. They aren't. Nor is it that biological variation related to sex does not matter. It does. It's just that not all humans fit neatly into the categories of female or male, and biological measures of human bodies rarely segregate into two non-overlapping categories. Neither "female" nor "male" describes a uniform or distinct biological type.

I will conclude by discussing why a binary view is a detrimental way to think, and talk, about sex biology and the human experience. Reproductive biology is an important structuring part of human lives; however, producing ova or sperm, having XX or XY chromosomes, or having a clitoris or a penis, does not tell us nearly as much biologically as many believe. Nor does it consistently or accurately inform us about an individual's childcare capacity, homemaking tendencies, interest in literature, engineering and math capabilities, or tendencies toward gossip, violence, compassion, or a love of sports. By contrast, placing reproductive biology in the context of the rest of the body, and in relation to behavior, history, society, and experience, we are much better prepared to ask, and answer, questions about health, habits, proclivities, happiness, and the many ways to successfully be human.

However, at its core, biology is about evolution, and evolutionarily speaking there is a lot of variation in sex biology and behavior, both across and within species. So, to really understand how biology and sex work in humans, we need to start not with us today but right back near the start of life on earth, with the evolution of sex.

1

The Evolution of Sex

IN THE BEGINNING . . . There was no sex. The earliest life forms on earth reproduced asexually, internally copying their key biological material and then dividing into two versions of themselves.[1] These organisms were, and remain, the prokaryotes, that is, microscopic single-celled organisms most people call bacteria (which scientists divide into archaea and bacteria)[2] and blue-green algae (cyanobacteria). Then, somewhere around two billion years ago, evolutionary changes resulted in what we call "protoeukaryotes" that evolved into the common ancestor of all eukaryotic organisms, including animals and plants. Eukaryotes have DNA in the form of chromosomes contained within a nucleus. Eukaryotes mostly reproduce via a process of combining genetic information from two different individuals to create a new individual, or, as that process is commonly called: sex.

The process by which sex evolved is complex but basically involved restructuring of the ways in which DNA is packaged and copied, the internal structuring of the cell's physiology, and the ways in which cells divide and fuse. A main challenge for sex is the ability to create a copy of one's genetic material and package it so that it can meet up and fuse with another of your same kind of organism's genetic material and create a new organism.

Prokaryotes could already copy their genetic material and divide. The trick was to keep the system of copying genetic material, but not divide into two new organisms, and create a little package of that material that could be combined with other such packages (called gametes) from other individuals.

Over hundreds of millions of years, single-celled eukaryotes developed the innovation of "mating types"—different forms of the same organism that could produce slightly different gametes. Once two different mating types got their gametes to fuse together, they created a new organism with two copies of all genetic material. Each organism produced via the fusion of gametes from two mating types is a little genetically different from each of the parents due to the mixing and matching of genetic sequences that happens during the gamete production and fusion processes.

From the very start of sex, variation was the name of the game. Even in their earliest appearances, the number of mating types (often called "sexes") per kind of species has been variable, ranging from two and sometimes three in most animals, to as many as seven in single-celled organisms, thirteen in slime molds, and thousands in some fungi. Sex across its evolutionary history is quite variable.[3]

Why Sex?

On the face of it, sex is a paradox. Wouldn't life be easier if organisms just kept copying themselves to reproduce? Why add extra challenges and potential problems? In 1975, a classic book on sex and evolution began with the comment, "This book is written from a conviction that the prevalence of sexual reproduction in higher plants and animals is inconsistent with evolutionary theory."[4] If sexual reproduction is more difficult and complicated than asexual reproduction, why did it evolve?

The favored hypothesis is that sex is a response to dealing with changing environments. Sex is a biological way to generate more variation for organisms such that they have better chances of meeting challenges the world throws at them. But sex a risky venture.[5]

Here is a simplified "sex is better than no sex" scenario . . .

Imagine a single-celled organism living in a pond, an amoeba-like thing that filters water for food. It might do just fine asexually copying itself if the water temperature stays more or less constant. But what if the water temperature shifts? The organism might not be able to deal with the new temperature. However, there could be a multitude of these organisms in the pond, each slightly different in its ability to adjust to fluctuating temperatures. Blending genetic material with another similar but distinct member of one's species could be a good option because it can offer greater flexibility and thereby a higher chance of survival for the offspring. But not all new genetic combinations created by sex do better. In fact, some do worse. Such is the risk of sex. It's the overall payoff that matters: if some sexually produced offspring do better, on average, compared to those of asexual reproducers, the system (sex) has a chance of catching on. And lucky for us it generally does. Otherwise, all life on earth would be asexual, and the world would be a lot less interesting.

Different Gametes = Biological Sex Differences?

Once sexual reproduction (the fusion of gametes) became possible, there was a quick transition from isogamy (making same-sized gametes by all individuals in a species) to anisogamy (making different-sized gametes by different individuals in a species) in many (but not all) kinds of life. The exact details of the emergence of anisogamy are not fully known or agreed on,[6] but it's coincident with the evolution of more complex forms of multicellular life over the last billion years. Today, having two different types of gametes is the norm among most sexually reproducing beings.[7]

Anisogamy usually manifests as two types of gametes: one large and one small. The larger gametes are called *ova* (colloquially known as "eggs") and the smaller gametes are *sperm*. Both types are haploid cells, meaning they each carry one copy of each parental chromosome, or 50 percent of the full complement of an organism's DNA (remember: all organisms produced by sex have two copies of the genetic material, one from each parent). However, ova also have the full range of cellular machinery and large assemblage of proteins and enzymes, while sperm only have a few additional proteins and enzymes accompanying the DNA. On average, each individual ovum is more energetically costly to produce than each individual sperm, due to the size and contents of each. However, it is not accurate to compare one ovum to one sperm in the sense of costs of production because in most sexual reproductive systems (especially mammals), many, many, many more sperm need to be produced than ova. For example, in humans, the correct comparison is 1 ovum = ~15 million sperm, as that's the number of sperm per ejaculate necessary for effective chances of a successful sperm-ovum fusion.

This difference in size and contents of gametes became a central focus in the nineteenth and twentieth centuries, when researchers began to hypothesize downstream differences in the mating types, or sexes, that produced them. One might think that it was the study of gametes that led the researchers to develop the hypotheses that gametes are the key to sex difference, but it was the other way around: belief that there had to be a key difference between the sexes—evidence of a biological binary—led to the focus on gametes.

The idea of biologically distinct sexes only became common in the eighteenth century. From the early Greek ideas about sex of Aristotle and Hippocrates and the Roman anatomical ideas of Galen through the Renaissance, and into the start of the eighteenth century, science and the medical world did not consider females and males as two separate kinds of biologies or beings. Rather, they were seen as hierarchically ranked (male above female) versions of the human form. This was termed the "one sex model."[8] But from the eighteenth into the nineteenth century, the belief in a "two sex" model, with males and females reflecting different biologies, emerged, but it had no specific definitional focus. It is from this newer two-sex worldview that the hypothesis of differences in gamete size as the key to female and male distinction emerged.[9] The general assumption, which became central for many biologists in the twentieth century, was that the two gamete-producing versions (sexes) of a given species have substantive biological differences in the production of their gametes and in the physiology and behavior associated with reproduction. Given these assumed biological differences, it stood to reason that each gamete-producing version (each sex) would be under distinct evolutionary pressures because of biological differences.

In 1871, Charles Darwin proposed the central evolutionary argument for differences between the sexes—that there are evolved differences in males and females and that they are based on different investment in reproduction. This argument was focused on anisogamy (different-sized gametes) by the biologist Angus Bateman in the late 1940s. Darwin[10] argued that evolutionary processes, called natural and sexual selection, made males and females very different, and Bateman[11] argued that the primary cause of this was "that females produce much fewer gametes than males" and that anisogamy would ultimately result in "an indiscriminatory eagerness in the males and a discriminatory passivity in the females," with "greater dependence of males for their fertility on frequency of inseminations." Basically, Bateman argued that cheap sperm and costly ova drive evolutionary processes, making males and females very different.[12] He developed his arguments by looking at flies where he thought males' energetically inexpensive sperm were their only "cost" for reproduction, and thus reproducing for them consisted of trying to get as much sperm into as many as females as possible. On the other hand, he saw the ova produced by females as biologically very costly, such that females were limited in how much they could reproduce and had to carefully consider each mating because they could only have a limited amount relative to males. Bateman's argument (cheap sperm and costly ova make males and females very different organisms) became baseline theory for evolutionary biology around 1966 and centralized in the world of assumed "biological fact" by another biologist, R. L. Trivers, in 1972.[13] Trivers connected anisogamy to parental care via a simple mathematical equation. Bateman's notions of sex differences entered near universal "truth" in biology with the work of E. O. Wilson in 1975 and G. A. Parker in 1979.[14]

As it turns out, Bateman was mostly wrong.

The problem came down to a counting error. Bateman made key errors in how he counted mating success for males and misrepresented the fruit fly mating system he was studying. More than thirty years of reexamination of Bateman's experiments, continued work on his model subjects (fruit flies), and in-depth study of sexually reproducing species demonstrate that acceptance of Bateman's paradigm as a "law" of biology led to two key problems. First, it led to a simplification of Bateman's data, resulting in an incomplete and biased understanding of his original findings. Second, because of the oversimplification of how reproductive biology works, it hampered scientists' ability to interpret reality with regard to male and female sexual behavior.[15]

This is a serious hitch in the science of reproduction. Bateman's early studies, and the authors that subsequently enshrined them into biological canon, have led to widespread confusion, mistaken assertions, and a deep-seated belief in anisogamy as the core explanation of everything about sex. That is not to say there aren't patterns of differential stressors and challenges across different reproductive systems. There are. Evolutionary histories have shaped sex biology into a range of typical patterns for ova and sperm producers, and we can learn a lot from the study of those patterns. However, while the gametes themselves mostly come in two types (a binary), most of the rest of the biology in organisms does not. Because they rely on the simplistic, and erroneous, assumptions about anisogamy, many of the classic assertions about *why* differences in sex biology emerge, *what* those differences are, and how we should think about sex, are incomplete and sometimes flat-out wrong.[16] We'll deal with this topic in greater detail in the following chapters. But before diving into the giant debate about sex biology and humans, we need to cover some basics, because reproductive biology is more than just gamete making.

Getting Gametes Together

The goal for any sexually reproductive system is to develop a method for ensuring that gametes get together, fuse, and prepare to develop into adult organisms. Not surprisingly, life on earth has developed a huge range of ways to get this done. Evolution and biology are anything but simple or uniform. While individuals in many species are large- *or* small-gamete producers for much or all of their lives, there are some species of animals where all members are simultaneously large- and small-gamete producers, and a number of species where individuals switch back and forth between being large- and small-gamete producers across their lifetimes. For most biologists, the term "female" describes an organism when it is a large-gamete producer, the term "male" when it is a small-gamete producer, and the term "intersex" is used for individuals that produce both types of gamete at the same time.[17] When this naming system is used across different species, some organisms would be called male or female or intersex their whole lives, but for others, the terms would switch depending on the life stage they are in. Be warned, however, because the biology and behavior of organisms sharing the terms "male," "female," and "intersex" are rarely identical within each category across different species. Take, for example, the following four female mammals: a queen mole rat, a dominant hyena, a doe kangaroo, and a mother cotton-top tamarin monkey. Their biology, behavior, and life experience, while sharing a few things in common, are all radically different from one another. In fact, the patterns and experiences of reproductive biology are not even necessarily the same even within individuals in a single species with the same reproductive organs, such as between a queen mole rat and a regular female mole rat.

Things are really complicated in reproductive biology, and the rest of the body, beyond the gametes, but an overview of the typical patterns in systems for gamete fusion is a necessary starting point in understanding sex biology. In the most basic reproductive systems, individuals produce gametes and eject them into the local environment (usually water), ideally in proximity to another member of their species. And they leave it to the gametes to find each other and fuse, creating a *zygote*.

This is called "spawning" and is common in aquatic species, including many fish, jellyfish, crabs, starfish, sea cucumbers, aquatic insects, and even some amphibians.

The next level of reproductive complexity involves one gamete-type producer (usually the large-gamete producer) creating some form of gel or other gooey/sticky matrix to deposit the ova in a cluster, which is often then attached to some object in the environment, like a plant or a rock in a pond, river, or ocean. Sometimes members of a given species carry the large sticky masses of ova on their bodies. Individuals who can produce small gametes approach the cluster of ova and eject sperm onto/into it, and thus fusion can take place, creating a mass of zygotes in the gooey matrix. A lot of amphibians, like salamanders and frogs, and many fish reproduce this way. Stepping up to the next level of complexity (physiologically at least) are systems wherein one sex develops an internal physiology that takes in and stores a cluster of already fused gametes (zygotes). Usually, this internal storing of the zygotes until they are ready to move on their own is done by large-gamete producers, but in a range of species, such as seahorses, it is the small-gamete producers that internally store and later "birth" the zygotes.

It is with internal fusion of the gametes that things get a bit trickier, as the small gametes need to get into the body of the large-gamete producer, reach the ova, and then the products of the fused gametes need to be excreted into the world at some later stage of development. There are more than a few twists to internal fusion. The first step is to get the gametes together in the body of the large-gamete producer, which usually involves some form of copulation (see below), and then there are a few variants on internal dynamics of the system. One way to deal with internal fusion and subsequent development is the production of a nutrient-rich, protective, solid, organic container, called an "egg." The development of the egg is catalyzed by the gamete fusion process and the contributions of additional physiological structures in the large-gamete producer who then excretes it. The zygote then develops inside the organic container but outside the body of the large-gamete producer.[18] The organic container (egg) system is found in many animals, from insects to reptiles to birds and even a few mammals (like the duck-billed platypus). Eggs are fantastically interesting structures, and the internal physiology associated with them is varied but has some core consistencies (described below).

The final type of internal gamete fusion and development system is the more evolutionarily recent, and highly complicated, mammalian innovation of gestation and lactation. Basically, in this system the large-gamete producer develops a specific internal physiology (a uterus) that enables it to act as a host for a process wherein a small number of large gametes are selectively released into a section of the internal biology. Then behavioral and physiological actions (copulating) requiring specialized organs (genitals[19]) occur between small- and large-gamete producers, and small gametes are introduced into the large-gamete producer's internal system. Then, if gamete fusion occurs, the resultant zygote implants inside the tissues of the large-gamete producer's uterus and develops into an embryo and then a fetus.

This is a relatively long-term developmental process that involves shared physiological connections between the developing zygote/embryo/fetus and the host large-gamete producer, with the embryo/fetus wholly dependent on, and part of, the host's body. At a certain point in development, the host excretes (births) the developed fetus (an *infant*), and some form of caretaking involving high-quality nutrition (lactation) ensues. Frequently, the host and/or other members of the species help the developing young individual for an extended period of time. This particular system sets up some expanded physiological obligations for large-gamete producers and throws in a whole new set of biological and behavioral complexities, and constraints, for the reproductive process.

Let's Talk about Gonads

The basic unit of reproductive biology is the gonad.[20] We all have them. This nifty little organ produces gametes. Gonads that produce large gametes (ova) are called ovaries, and those that produce the small gametes (sperm) are testes. In most species of animal, a majority of individuals usually have one of the two gonad types (ovaries or testes) as adults. However, there are a number of species in which all individuals shift from one gonad type to the other across their lifetimes, others that have both gonad types as adults, and there are also a number of species (largely invertebrates) that have single gonads (called ovotestes) that produce both types of gamete. There is also significant variation within species, including humans, where a small but not insignificant percentage of individuals have some mix of the above (an untypical gonad, two gonad types in the same body, or even ovotestes). Most gonads also have associated duct and tube systems for the storage and excretion of the gametes, a significant/crucial component of structuring a species' sex biology.

In vertebrates, gonads also produce a range of hormones related to the growth and functioning of reproductive tracts, accessory sex glands, and organs related to copulation. These hormones contribute to a range of behavioral actions associated with copulation and other facets of development and functioning in organisms' bodies. Testes and ovaries emerge from the same embryonic tissues, but they vary in certain aspects of their mature physiology. All types of gonads produce the same hormones, but they often do so in differing levels. Therefore, individual organisms with either testes or ovaries, individuals having both, and those with ovotestes differ with respect to levels of some hormones (but not kinds) between them.

Details of Internal Fusion Systems

For animals in which gamete fusion is internal there are a variety of additional organs and physiological systems that make up key parts of sex biology. First, there are copulatory organs whose role is to facilitate the excretion and/or absorption of gametes. At a basic level, these are a variety of bodily modifications for clasping during copulation and simple openings connected to gonad duct systems. At a more complex level, they include invagination of body surfaces and cavities, external genitals (such as a clitoris and penis), and internal structures (such as a vagina) that interact to facilitate getting the large and small gametes into interaction with one another.

As for the core layout of reproductive organs, most vertebrates have a cloaca, a common opening/chamber for waste excretion from the urinary and digestive tracts that also serves as the exit point for sperm and/or ova. In most birds and reptiles, the cloaca is divided into a urodeum for the urine, a coprodeum for fecal matter, and the proctodeum, which is the final chamber next to the cloacal opening associated with copulatory organs and the excretion of sperm and eggs (not ova but actual eggs).

In vertebrates that lay eggs, there is specific differentiation in internal sex biology between large- and small-gamete producers. For example, egg-producing reproductive tracts develop into a system whereby the ovaries produce ova that are transported to an oviduct, a long tube with multiple cell types across different parts of it. As the ova passes through the tube, they fuse with sperm (if present), acquire various proteins and materials that make up the yolk and white of the egg, and then stop in the shell gland. In the shell gland, the internal materials of the egg finalize, and the hard calcium-rich exterior develops around to contain them, after which the egg is excreted via the cloaca.

Mammals have even more complicated reproductive physiology. Most have some form of external genitalia derived from the embryonic tissues called the genital tubercle. In humans, typically, this tissue mass develops into a penis or clitoris, with a wide range of variation in the adult endpoints (see chapter 5). Associated areas of tissue (the labioscrotal/urogenital fold) develop into the scrotum or the labia majora, and other tissues of the urogenital folds develop into the skin of the penis and the internal labia (if present). Many mammals have internal testes, but several of them (including humans) have an external scrotal sack and testes that reside in it. All ovaries are internal.

A core difference between mammals and other animals is that mammals gestate. Zygotes develop inside the uterus into embryos and then fetuses.[21] To accomplish this, mammals evolved a complex system involving connection from the ovaries to a fallopian tube and into the uterus. In one lineage of mammals, including us, the uterus develops a special organ called the placenta plugging the zygote directly into the maternal physiology and providing nutrients and removing waste. This intense placental linkage between maternal and fetal bodies is distinctive in the animal kingdom. It creates a suite of patterns that structure the bodies of individuals with this reproductive anatomy relative to individuals who do not have it.[22] At the end of gestation, one or more fetuses are birthed needing to be fed a high-quality nutrient (milk) until they are ready to fend for themselves food-wise. To accomplish this, mammals develop mammary glands that produce high-quality milk during lactation. The ovary-fallopian tube-uterus physiology and mammary glands are typical of large-gamete-producing mammalian bodies, but not small-gamete-producing ones.

Caring for Infants Matters

Animal bodies are shaped by sex biology, and in many cases, their behavior is too. While reproduction for many animals ends with the excretion of gametes, there are others for whom care of the embryos, fetuses, and postbirth young is a central part of the reproductive cycle and shapes the animals' ways of living.

On the simplest end of the caretaking spectrum, in many species of fish and amphibian, adults will guard masses of zygotes/eggs and carry and protect masses of embryos ("brood" them) in their mouths, on their backs, or in invaginations or pouches in their body until the embryos develop into young capable of their own movement and feeding. Many reptiles stay with laid eggs and protect them until they hatch, with some snakes and lizards carrying the eggs internally, and then give "birth" to the hatchlings. Some reptiles, like crocodilians, even care for the young for a period after birth. Birds almost always care extensively for eggs and thereafter the hatched young. Bird parents provide warmth and protection to the eggs, then care and food for the developing hatchlings, and ultimately some level of training for adult life. And, of course, mammals take the cake in terms of pre- and postnatal care. Some lineages, such as primates, elephants, cetaceans (whales and dolphins), and many social carnivores, intensively care for young for years after birth.

For most of the nineteenth and twentieth centuries, biologists mostly assumed that because of the supposed implications of anisogamy, the large-gamete producers ("females") evolved as the primary caregivers and thus their bodies, and brains, were "designed" for parenting. In contrast, small-gamete producers ("males") were considered to not be under such pressures and would thus be shaped by evolution to maximize getting their sperm to the ova in order to enable the fusion of gametes, regardless of what happened postfusion. This simplistic view of anisogamy is wrong, and so were many of the nineteenth- and twentieth-century biologists' assumptions both about the limitations on "females"[23] and about parental care and its evolution. To be fair, from very early on, it was clear to many scientists that there was indeed variation in parenting and other reproductive patterns, and sex-related biology, across the animal kingdom, but these variations were often seen as exceptions and abnormalities, and studied as such. However, as we'll see throughout this book, what were previously thought of as exceptions or minority patterns are more often than not a core part of the actual range of biological variation in and between species, and seeing and studying them as part of an expected range offers a better understanding of biology and behavior.

Clearly, there are some biological patterns that structure large-gamete-producing bodies differently than small-gamete-producing bodies, especially in those species that lay eggs or gestate and lactate. However, these differences in reproductive anatomy do not predict the entirety of the animal's biology, behavior, and ecology. In fact, when it comes to care, there is much variation, and the specifics of that variation are central to structuring how any given sex-biology system works. The assumption that large-gamete producers are uniformly destined to be the caretakers of young is not true.[24] In fish, "males" offer care more often than "females."[25] Among amphibians, there is a wide array of caretaking patterns across all gamete-producing types,[26] and with birds, both gamete-producing types are joint primary caretakers in nest brooding and posthatching care in 81 percent of all species.[27] When it comes to mammals, reproductive biology dictates a specific postbirth role of lactation for large-gamete producers, but regarding all other forms of active caretaking of infants and juveniles, there is a range of variation from small-gamete producers who do 90 percent of the infant care, to fully cooperative care by an entire group, to exclusive maternal care. And this range of different patterns generates a diversity of mammalian sex biology and behavior between, across, and within gamete-producing categories.[28]

There Are Many Ways to Do Sex Biology

At its core, the fusion of gametes forms a zygote, then an embryo, then a fetus, and eventually an adult organism. Thus, any biological system dealing with sex must be prepared to respond to three core challenges. The first is to make gametes, the second is to get those gametes together, and, finally, the last is to provide a physiologically and ecologically viable environment for a zygote to develop and mature.

The first two challenges create specific biological characteristics of organisms: gamete-production physiology and gamete-placement biology and behavior. The third challenge, finding a viable environment, ranges from simply finding a spot in the water of the right temperature and excreting gametes, to internal gamete fusion and laying eggs, to a range of internal gestation strategies, to complex parenting behavior. Clearly, this varies by type of organism. For many, some form of investment and/or care beyond gamete fusion is needed.

When parental care is needed, a new set of challenges emerges. There must be a physiology sufficient to facilitate the development of the zygote into an embryo, and then into a young organism. Then behavior sufficient to facilitate care for the development of the young organism(s) is usually required, sometimes for many years after birth. Finally, animals that have the physiology and behavior to care for young also need a socioecological system (a way of being socially) and set of responses to ecological challenges in which their particular physiology and behavior can be effective. It is in this physiology and behavior of parenting that the dynamic variation in sex biology becomes most apparent and most interesting for a conversation that is eventually about humans. The ways in which organisms interact with their environments and each other are important in understanding sex biology.[29]

Surveying the range of animal sex biology, a group of biologists recently summarized their findings as follows: "'Sex' comprises multiple traits. . . . Individuals may possess different combinations of chromosome type, gamete size, hormone level, morphology, and social roles, which do not always align in female and male specific ways or persist across an organism's lifespan. Reliance on strict binary categories of sex fails to accurately capture the diverse and nuanced nature of sex."[30] In short, there are many, many ways to "do" sex biology effectively. The next chapters lay this out for other animals and then for humans.

2

Animal Sex Biology

MIXING IT UP

EVERY TEXTBOOK lays out how animals are supposed to "do" sex biology. But most animals don't read textbooks.

Sex biology is often misunderstood as a story of two kinds of creatures, the male and the female. For example, many assume male mammals are larger than female mammals. But there are multiple examples of large-gamete-producing individuals that are the same size as—or larger than—small-gamete producers.[1] And in many species, including humans, there is huge variation, and overlap, in size across populations and between individuals within the large- and small-gamete-producing categories. Most people also incorrectly assume that female animals are less aggressive, provide most or all care of young, and are choosier in mating than males, when counterexamples exist among species of fish, birds, lizards, and mammals.[2] The assumption of fixed "sex" differences and "sex" uniformity either between or within species is a major stumbling block to understanding biology and behavior in organisms.

Science is supposed to constantly adapt to new data and analyses. The first periodic table of elements had 63 entries (in

1869); today, we have 118 due to discoveries of new elements. A century ago, tool use was seen as the near-exclusive domain of humans, but the study of apes, monkeys, birds, elephants, and many other animals revealed tool use across much of the animal kingdom. The longer a system, a species, or a group of animals is studied, the more fine-tuned our understanding becomes.

Of course, there are patterns and trends associated with reproductive systems that are common across much of life or in specific lineages. But there are also numerous variations and modifications in the patterns of sex biology in animals. In some lineages, and sometimes within the same gamete-producing categories, many previously considered "atypical" variations are as common as the expected patterns. In others they are not. Identifying what is "typical" or common and assuming that this identification represents sufficient understanding of a species or system is a scientific error. Ignoring biological variation, even if small, is bad science because the goal of biology is understanding how bodies are made and work: how all the parts relate to one another. So, ignoring aspects of some parts and the variation of those parts blinds us to key data needed to do the best science. Authentic understanding of the world grows as more of the range of what is really out there is added in.

Large- and small-gamete producers can and do experience different evolutionary pressures resulting in some sexual selection for sex-biology "roles." But any one pattern, even if it's very common, is seldom ubiquitous, and the actual biological outcomes—what the "roles" and bodies look like and do—vary substantially across different species and lineages.[3] In mammals, the reproductive physiology associated with gestation and lactation shapes the bodies of those who gestate and lactate relative to the bodies of those who do not. However, even in mammals with this very specific set of physiological constraints, how this system translates to the specifics of body size and shape, hormone patterns, care of young, and sexual behavior in large-gamete producers alone can vary dramatically from opossums to rats to hyenas to dolphins to monkeys to elephants to humans.[4] So, while there are general sex-biology patterns that are common, even typical, it is in examining the variations across biologies, within and between kinds of animal, that we see there is not one "best" or "right" or single way to characterize sex biology in the animal kingdom.

Let's take a brief tour.

The Hymenoptera (bees, ants, and wasps) are haplodiploid, meaning they have a genetic system in which some individuals develop from fused (diploid) zygotes and others from unfused (haploid) sex cells. But this is not a binary "male/female" system. In bees, the system of sex biology includes two types of genetic systems that produce three types of bodies: one small-gamete producer and two large-gamete producing physiologies. But one of the large-gamete types does not actually produce any gametes (usually) or engage in sexual activity, while the other one can and does both.[5] So instead of simply labeling both types "female," we call the first type "workers" and the second type a "queen." The small-gamete producers are called "drones." The queen is much larger than the other two body types and can control the pattern of gametic fusion in her own reproductive tract to manage the ratios of the three body types in subsequent generations. The behavioral, ecological, and physiological lives of bees involve three sex-biology types.

In the Hemiptera (the official name for about eight thousand varieties of wingless, sucking insects commonly called scale insects), all zygotes are diploid, but in order to develop into a small-gamete producer, the zygote must undergo a "turning off" of chromosomes or lose certain chromosomes right after gamete fusion. In other insect lineages, all offspring of any given individual will be only large-gamete or only small-gamete producers. Yet in others, which reproductive physiology (small- or large-gamete producing) will develop is entirely temperature-dependent. Parthenogenesis—asexual reproduction by which large gametes develop into zygotes without any need for small gametes—is also found in a minority of insects. Strangely, while bodies with both small- and large-gamete-producing physiology are fairly common in many invertebrate animals, they are extremely rare in insects.[6] Small- and large-gamete producers' bodies, behaviors, ecologies, and physiologies are not uniform across the insect world; there is no one universal reproductive biology, or one consistent way to become or to be "female" or "male" in insects.

Worms, too, have fascinating sex biologies. Earthworms, for example, are all intersex. As adults, each individual has fully developed and functioning testes and ovaries along with a series of organs and glands that act to bring large and small gametes together, either in the individual's own body or via a mutual gamete exchange with another earthworm. Additionally, certain species are also capable of parthenogenesis (asexual reproduction with no gamete fusion at all).[7] The sex biology of earthworms enables a variety of sex "roles" but is not characterizable as a binary. For example, in some species, the same worm can have "self-sex" between the two reproductive physiologies in their bodies, have sex with another individual involving the mutual exchange of both large and small gametes, and undergo parthenogenesis. That's pretty much the whole range of reproductive options for living things in one individual. Alongside this dynamic in sex "roles," the development of such worm sex biology also involves pretty complex sex-determination systems. Most worms have polyploidy,[8] with three or as many as twelve (!) sets of chromosomes related to sex biology.

Other types of worms have different variants of sex biology. The roundworm *C. elegans* has only two sex-biology types, but they are not the typical pattern of a small-gamete producer and a large-gamete producer. In *C. elegans*, individuals usually either produce both gamete types or just small gametes. The large-gamete-only sex biology rarely shows up. However, individual *C. elegans* can flexibly alter temporarily, or permanently, the specifics of their reproductive system (that is, go from one "mode" of gamete-producing physiology to another) in the face of environmental challenges.[9]

Organisms with backbones offer yet a new set of variations to consider. Take teleost fish, which are the most common type of bony fish. They have a sex biology that is flexible, with major changes in reproductive physiology within the lifespan of individuals being common. Many fish are "sequentially intersex," meaning individuals start out with one set of sex biology and then change one or more times to different sets of sex biology across their lives.[10] Such changes involve transformations, as adults, of multiple biological systems, and include behavioral, anatomical, physiological, neuroendocrine, and molecular processes.[11] Fascinatingly, the bluehead wrasse usually develops into an adult as a large-gamete producer and then might shift to a small-gamete producer if their body size grows very large. But this usually only happens when the largest bluehead wrasse in a small area of a reef dies. When this happens, the next largest wrasse changes their sex biology and shifts from large- to small-gamete-producing physiology and takes over as the sole small-gamete producer in that specific area of the reef. However, the system is not without its variations. Sometimes a small, younger wrasse does not mature into a large-gamete producer but has a body that looks just like one, even though it is producing small gametes. These individuals are called "female-mimics" and covertly mate with other small-bodied bluehead wrasses without the large local "male" noticing them.[12] The bluehead wrasse has three sex-biology types, and any individual in the species may or may not "be" one, two, or all three of them during their lifetimes.

Other fish species switch in the opposite direction, starting out as a small-gamete producer and switching to a large-gamete producer only when they become much larger bodied. Many anemonefish ("Nemo" from the Disney movie, for example) live in groups of two dominant reproducing adults and a collection of nonbreeding subordinate individuals. The larger bodied of the breeding pair produces large gametes and the smaller one makes small gametes. But, if the larger one dies, then the smaller member of the breeding pair undergoes a sex-biology change to become a large-gamete producer, and one of the nonbreeding members will step in to take over as the small-bodied "male" of the breeding pair.

All California sheepshead fish change from large- to small-gamete-producing physiology once they grow large enough to defend a territory.[13] But some species change multiple times. Okinawa pygmy gobies start out as a large-gamete producers, and then, if they grow to become the largest goby in the group, they transition to small-gamete-producing physiology. However, if a larger fish enters the group, the one that just switched from large- to small-gamete producing switches back to large gametes in the presence of the new, larger goby.[14] The triggers for sex-biology change in fish are frequently social (a change in makeup of the social group or local breeding population) or directly tied to body size. In some coral gobies, change in sex biology in one direction or the other depends on the sex-biology makeup of the population they are born into or migrate into. The gobies can shift their biology in either direction (toward large- or small-gamete-producing physiology) so that any two fish can form a breeding pair. The ability to shift sex biology flexibly by all individuals in a population reduces the costs and risks incurred by moving between the coral clusters where the gobies like to live.[15]

Internal gamete fusion is a whole different ballgame from the external fertilization we have just learned about in fish. Having the gametes fuse internally is associated with a set of specific reproductive organs in birds, reptiles, and mammals. With internal fusion in reptiles and birds, the large-gamete-producing individuals' bodies develop egg-producing physiology and then excrete the egg. In mammals with internal gestation (development of the zygote into embryo and then fetus all internally), there is need for a very distinctive physiology, including fallopian tubes, uterus, and, in most, a placenta as well. These specific organs and bodily modifications in reptiles, birds, and mammals create patterns of differences in development and adult physiology between large- and small-gamete producers, giving them reproductive physiologies that are less flexible than those in the invertebrates and in fish. It is in these birds, reptiles, and especially mammals, where one sees, on average, greater physiological differences between the sex biologies of large- and small-gamete producers. However, what those patterns are across and within species, what is "typical" for any given reproductive physiology, and how much overlap exists between the bodies and behavior of large- and small-gamete producers within a given species vary quite a lot.

Take reptiles, for example. They have a distinctive climate-influenced sex biology: many species have a temperature-dependent sex determination.[16] In many species of lizard, turtles, and crocodilians, the sex biology of a developing embryo is not predetermined by a specific genetic makeup. Rather, it's a suite of hormonal responses driven by the local temperature around the eggs during a critical period of development that shapes the reproductive tracts and eventual associated adult physiology. For many reptiles, developing large- or small-gamete-producing physiology is a context-dependent outcome not predetermined by the genetic results of gamete fusion.

It's not just how sex is determined that is interesting in reptiles; patterns of sex biology can also vary greatly across species. There are species where females are larger than males (in sea turtles and some snakes), others where males are larger than females (in many lizards, such as Gila monsters where small-gamete producers have larger heads), and yet others where there is little or no size difference (such as grass lizards). There are a number of hypotheses for why these diverse patterns emerge in different reptile species, but it's likely that both competition between small-gamete producers (likely due to sexual selection) and environmental constraints on body size (for both small- and large-gamete producers) are involved.[17]

There is a particularly interesting mode of sex biology in a cluster of lizard species (and some snakes): there are no small-gamete producers.[18] In some whiptail lizard species, small-gamete-producing physiologies have disappeared, and only large-gamete producers with egg-laying reproductive tracts remain. These lizards evolved a complicated way to duplicate their chromosomes and divide the results so that when they produce large gametes, there is already a full complement of genetic material in them with novel rearranging so it's not just cloning. These gametes have no fusion with other gametes but are ready to develop into an egg and eventually into a new individual.[19] In these lizards, sex biology not only isn't binary, it's unitary.

Finally, reptiles, regardless which sex-biology variants they exhibit, offer little in the way of care of the young. Some species do guard eggs before they hatch, and other species carry the eggs in their cloaca until they hatch rather than laying them, but their care goes no further. Crocodilians are the only exceptions, with adults doing some caretaking after the eggs hatch.[20] It's in the animals that go all-out for their offspring posthatching (birds) and postbirth (mammals) that a whole new level of sex-biology complexity emerges.

Birds lay very large eggs, relative to adult body size, that hatch into young who cannot fend for themselves (this type of offspring is called "altricial"). Birds must care for their young. In most birds, parental care is overseen by both the large- and small-gamete producers. Such care likely emerged in protobirds (better known as dinosaurs) from some forms of early care of young that increased and facilitated the development of very large eggs and highly altricial young.[21] While most birds have biparental care of young, the details of that caregiving vary widely. How large- and small-gamete producers contribute to such care is related to differences in social systems, mating systems, environmental context, and bodies.[22] Parental care is central to bird existence, but across—and occasionally within—species, there is not a consistent one-to-one correlation between a bird's reproductive physiology (being a large- or small-gamete producer) and how it participates in care for young once the eggs are laid and after the young hatch.

In many bird species, there is sexual dimorphism (differences in the shapes, sizes, and/or colors of bodies) between large- and small-gamete producers, but what form that takes is not consistent or in a single pattern across all species. Body-size differences between large- and small-gamete producers often relate to aspects of social and mating systems, to the patterns of parental care, and to aspects of the specific ecologies of that species. Like so many other animals, the patterns of how sex biology varies with bird species is dynamic and multifactorial: there are not simply two uniform sex types.[23] For example, in many species, small-gamete producers are larger bodied. But in some, especially birds of prey (like hawks and eagles), large-gamete producers are larger bodied. In birds where small-gamete producers do most of the competing for mates, they are usually more colorful or have the most complex songs, but not always. Also, sometimes there is intense competition between small-gamete producers and almost no sexual dimorphism. And in some species (like kingfishers and some parrots), large-gamete producers are more brightly colored than small-gamete producers. In songbirds, small-gamete producers sing most. But in some species, large-gamete producers also sing, and in those species, they are often more colorful.[24] There are even some bird species with multiple "sex" categories. The white-throated sparrow has some changes to its chromosomes that effectively produce four chromosomal types that have different plumages and a mating system wherein certain types are not compatible with others. So, while there are only two gamete-producing physiologies in the species, there are functionally four sexes in the reality of the actual mating system.[25]

The takeaway is that while there are typical trends in bird morphology and behavior that differ between large- and small-gamete producers, especially in that only large-gamete producers lay eggs, there is no one-to-one universal correlation between a bird's reproductive physiology and its body size, plumage color or brightness, mating or parenting behavior, song patterns or the specifics of its social system. "Sexes" are not uniform things in birds.

As varied as birds are, mammals, including humans, do sex biology differently from everyone else. The bodily processes associated with mammalian internal gametic fusion and the potential for, and actuality of, gestation result in some very specific physiological patterns. Mammals also produce and provide high-quality nutrition to infant(s) during the earliest stages of life. Large-gamete-producing mammals typically can gestate and lactate. Small-gamete-producing mammals cannot gestate (with some variation in intersex individuals, including in humans) and rarely lactate.[26] Mammals also exhibit high levels and a wide variety of postbirth care of their young, including care by one adult, care by two adults, care by many adults, and care by entire groups. It is common, but not at all ubiquitous or uniform, that large-gamete producers are the primary caretakers relative to other group members.

While sexually dimorphic patterns associated with reproduction are common in mammals, they are neither ubiquitously present nor consistent across all species. There is variation in body shape and size, physiological processes, hormone patterns, genitals, caretaking, and a range of behavior. While one can certainly assert that there are typical patterns of reproductive physiology and behavior in mammals, there is simultaneously a wide range of variation rejecting the assertion of a single "way to be" within and between mammalian species.[27] There are many successful ways to be reproductively effective mammals. Hyenas, "monogamous" primates, and naked mole rats offer fascinating examples.

Unlike the devious characters in the *Lion King* or the scoundrels or scavengers often represented in books and nature documentaries, spotted hyenas are active and successful hunters, form strong and deeply bonded social groups, and mess with most peoples' assumptions about mammalian patterns of sex biology.[28] Spotted hyena large-gamete producers are larger, heavier, and more aggressive than small-gamete producers, and socially dominant over them. The external genitalia of these hyenas include labia that develop into what looks like a scrotum and a clitoris that develops into what is called a "pseudopenis" that is fully erectile and through which mating and birth take place.[29] One might think that such differences simply mean that hyenas with ovaries are more "masculine" than those with testes. That is to say they get more androgen hormone exposure during development and thus mimic "typical male mammal" development. But that's not the case. The patterns of androgen concentrations (those sets of hormones often associated with development of "typical" sex biology for mammals with testes) in spotted hyenas are complex and vary by and are dependent on age, whether animals reside in the group they were born into or not, their social status within the group, and on reproductive state.[30] Fetal large-gamete-producing hyenas are not "masculinized" by hormones in the womb. Case in point, the development of the genitals in hyenas with ovaries is not, as researchers once thought, controlled by testosterone or other androgen exposure but is, in fact, mediated by a very complex interplay between a range of multiple hormones. This distinct physiology is important because hyenas are not "intersexed" in any manner or form, or by any biological definition. They are neither "hermaphroditic" or "masculinized females." The hyena system reflects typical genetic and internal reproductive physiology patterns for large-gamete-producing mammals but demonstrates a nontypical way to be a successful large-gamete-producing mammal. Hyenas force a reevaluation of how one can, or should, define "masculine" or "feminine" in the context of sexual biology and motivate reflection about the value of simple models of "male" and "female" in mammals.[31]

Similarly, there is a whole group of primates that challenges the expected roles of large-and small-gamete producers. Most primates live in big groups of several males and females and young, and others live in groups of one male and many females and young. But there is also a range of species that live in small groups of two adults and their offspring. These so-called "monogamous primates" include the nighttime dwelling owl monkeys, the titi monkeys, the marmoset and tamarin monkeys of South America, and the gibbons (a small ape) of Southeast Asia. All are assumed to have a particular pattern of sex biology and behavior that differentiates them from the assumed mammalian pattern of large males and smaller, infant-focused females. The general argument is that these primates have a "monogamy package" of bodies and behavior consisting of specific deviations from the mammalian typical pattern. The variations on the typical mammalian pattern include large- and small-gamete producers of the same size and shape (monomorphic), small-gamete producers who do much or sometimes all of the infant caretaking, and no dominance differences across females and males. This "monogamy package," while deviating from the typical expected pattern for mammals, is still based on the assumptions of anisogamy, and thus still fits into the typical assumptions about mammalian sex biology. The problem is that this monogamy package does not exist.[32] Basically, the only thing that all of the "monogamous" primates share in common is that they primarily live in groups of two adults and some young. In some species that live in pairs, sex is only between the two adults (sexual monogamy). But in others, sex can happen with individuals outside the two-adult group (social monogamy). In some pair-living species, small-gamete producers are larger than large-gamete ones and do little to no infant care, as in many other nonmonogamous primates. In others, both types of gamete producers are the same size, or large-gamete producers might even be a bit bigger, and small-gamete producers do at least half if not 100 percent of the carrying and social care of infants. In pair-living primate species, the two adults are sometimes codominant, and other times the large-gamete producer is dominant, and there are also, occasionally, more than two adults in the group ("monogamous" species are sometimes neither monogamous nor pair living). Finally, the hormonal profiles across large- and small-gamete producers in the wide range of "monogamous" primates vary depending on their reproductive condition/status, whether they care for infants, the demographic makeup of the group, age, and dominance status. Even within what was assumed to be a straightforward "monogamy exception" to the typical mammalian system of sex biology and behavior, there is a huge amount of variation in the biology and behavior of these primates living in small groups.

Naked mole rats are particularly strange mammals. They have little hair, spend almost all their time in extensive burrows, don't regulate their temperature well, and live in large colonies with one reproducing, often tyrannical, "queen," and a few reproducing small-gamete producers, with all other members functioning as nonreproducing "workers."[33] The naked mole rat social system is superficially like that of some social insects (such as bees). However, there are no differences in sex biology regarding chromosomes or internal reproductive physiology between the two "types" of large-gametes producers ("queen" vs. non-queen). There are no biologically distinct "workers" as in bees. Nor is there pheromonal control of "workers" or any lasting biological changes that alter their internal physiology. The mole rat "queen" (sometimes there are two in a colony) has a longer body than other mole rats and behaves differently. While there are some hormonal differences in naked mole rats relative to other species, they do not show any of the massive sex-biology differences between sex-biology types that we see in insects. In other words, naked mole rats have developed an insect-like social and behavioral system without major modification to the standard set of mammalian sex biology. Naked mole rats demonstrate that expectations of a consistent pattern of social organization and behavior based on standard mammalian sex biology are not accurate.

Hyenas, naked mole rats, and "monogamous" primates are only a few examples of variation relative to typical mammalian patterns, but they illustrate that sex biology and its relation to bodies and lives is not a clear-cut process. Internal gestation, lactation, and care of young *does not* limit mammals to one type of social system, one strict set of behaviors, or one single way to successfully be "female" or "male."[34]

Is There a Sex Binary in Other Animals?

Chapter 11 of one of my favorite books[35] is titled "Beyond the Binary: Evolution's Rainbow," and it opens with a quote attributed to the twentieth-century evolutionary biologist J. B. S. Haldane: "*The universe is not only queerer than we suppose, it is queerer than we can suppose.*" Here "queer" means strange, fascinating, different, confounding of expectations. This quote is fitting for the reality for sex biology. Gametes (sperm and ova) in most animals can be described as binary, one large and one small. But bodies, physiology, and behavior are not so easily classified, and are queer indeed.

The distribution of large- and small-gamete-producing physiologies, what forms they take, whether they are found in the same body, different bodies, one body at different times, or even consistently within a given animal lineage varies. What "female" or "male" means physiologically, behaviorally, and ecologically can be variable across and within species, and a simple take is often insufficient to categorize, or understand, sex biology. Therefore, to effectively understand sex biology and its relation to bodies and behaviors, one must examine closely the evolutionary and ecological history, as well as the current biology and behavior, of a species or lineage. And nowhere is this more evident, important, and convoluted than in humans.

3

Humans Are Messy

ALMOST FIVE THOUSAND years ago, a group of people in the eastern Iberian Peninsula (near where Valencia, Spain, is now) laid to rest one of their kin and buried ivory, weapons, crystals, and ostrich egg shells alongside the bones. When archeologists discovered and analyzed the grave site, they asserted that the shape of the pelvic bones was typical of a male and the treasures indicated the individual was a revered leader. Further, because of the bone shape and the conclusion that the individual was a leader, they concluded the individual must be a man.

Fifteen years after the initial discovery, researchers analyzed the teeth using a new technique that detects proteins specific to X and Y chromosomes. Lo and behold, the revered leader's proteins reflected no Y chromosomes. The individual was reclassified as female.[1] Oops.

What happened here? The initial researchers assumed men's bodies and lives are one way and women's another. In this instance, they believed that the specific grave goods were signs of high social status and reflected a specific set of "men's" and "women's" social roles. And, in combination with the pelvic bones looking "typically male," they made the assessment to classify both the biology and gender and call the individual a man. But we don't know if this individual was classified as a "man," a "woman," another category, or even if those categories were the ones used in that society five thousand years ago. What we can say is that the individual had elite grave goods and a body the reflected a variation on what is often assumed to be "typical" sex biology. Placing contemporary assumptions about gender roles and a binary sex biology classification on past humans, or on present humans, is almost aways shoddy science. The moral of this story? Human sex and gender are complicated. Every human body, and life, is a blend of biological and cultural processes, and nowhere is that more evident than sex biology and the human experience. And that blend is what this book is ultimately about.

Human Sex Biology Is Complicated

In the big picture of human history, and in the biological sciences, the binary definition of sex biology is a recent invention. It emerged out of particular cultural and philosophical commitments centered in the European enlightenment and infused the biological sciences across the eighteenth to the twentieth centuries.[2] For at least the last century and a half, biologists generally use the terms "female" and "male" when referring to what type of gametes (ova or sperm) a sexually reproducing organism produces. But gametes don't magically appear out of thin air. They are created, managed, and maintained by complex physiological processes and organ systems. Biologically, "sex" involves all the processes of sexual reproduction—not just gametes.

Late in the twentieth century, scholars challenged the absolutism of the binary view of sex biology and proposed definitions of sex as bimodal (having two broad ranges that overlap extensively). Now, in the twenty-first century, many researchers see the bimodal model as still too much of an oversimplification. What is referred to as "sex" biologically comprises multiple traits and processes with variable distributions and patterns. The best current data and analyses produce a contemporary understanding that a "reliance on strict binary categories of sex fails to accurately capture the diverse and nuanced nature of sex."[3]

There are few universally consistent connections between the ability to make sperm, ova, both or neither, and the majority of bodily functions. Even in mammals, with internal gestation and lactation, there are variable connections between gamete-production physiology and bodily characteristics like body-fat, muscle density, height, blood pressure, metabolism, brain function, and a range of other aspects of biological function both across and sometimes within species. There are even fewer uniform connections between gamete production and specific behavioral patterns, aside from the specific behavior associated with giving birth and lactating. And even in the arena of birthing, the postbirth behavior and the investment by adults into raising offspring is not consistently associated with a specific gamete type. While one can describe gametes as binary (of two distinct kinds), the systems, behaviors, and bodies that produce them are not. And in the case of humans, it's even more complicated.

Humans Are Biocultural

The material structures of human bodies, as with all animals, are the product of complex interactions of biological, developmental, and environmental systems. Things like hair color and type, nose, feet, and head shape, the length of legs and arms, and the shades of our skin are generally described as our "biology." However, these structural parts of the body exist within and are part of the human cultural context. Humans shave, cut, or color their hair, bind their feet in shoes altering their shape, change their noses surgically, with ornaments, and via life experience, eat a range of foods, and participate in activities that affect growth patterns and limb morphology; they even alter the shades of their skin by activity, clothing, chemistry, and tattoos. The human foot may be a collection of tissues, ligaments, bones, and muscles arising from the interactions of genetics and development, but for every human, their foot—including its form and function—is as much shaped by their cultural selves as it is part of their biological structure.

How humans taste food, whom they are attracted to, the sports they play, and how well they play them, are always affected by biology and culture. Taste bud reactivity, muscles and coordination, and the targets of sexual desire all emerge from the mutual, and interactive, development of bodies and lived experiences. Adult height and weight, one's ability to score well on standardized exams, views on raising children, resistance (or lack thereof) to disease-causing bacteria, and even what any given person thinks of as natural behavior for a man or a woman, are a product of dynamic relationships that interweave the biological and the cultural, as well as the historical, into a single result: us.[4] Although we are biological organisms, the totality of the human experience cannot be reduced to either specific innate (biological) or external (environmental/cultural) influences. It is a synthesis of both: humans are biocultural.

The human brain is about 40 percent of its adult size at birth, relatively smaller than any other mammal. Most human brain growth happens out in the world, not in the womb. The brain and the entire nervous system develop, create, and alter connections and pathways and flows of biochemical signals in constant exchange with bodies, senses, the social and physical environments each human inhabits, including their own thoughts, perceptions, and experiences of all these things.[5] This is the embodiment of culture: human bodies are literally cultural organisms. Cultural experiences are integrated into skin, muscles, nerve fibers, neurons, and other physiological systems, and the entire process is central to the development, makeup, and function of the human animal.[6]

Humans are born into a world of social and physical ecologies, patterns, institutions, and ideologies that become inextricably entangled with our biology, even before we leave the womb.[7] Even basic perceptions such as smell, color, sound, and taste are mutually shaped by physiology and cultural experience. Think about many of the things associated with sex biology in humans: femininity, masculinity, sexuality, sports abilities, child-care, health, body shape, how one speaks, how one walks, and so on. Every single one of these elements reflects a deep interweaving of biological, cultural, and social factors, and more. Every aspect of the human experience of sex biology is deeply biocultural. And gender is central in that experience.

An Ova Is Not a Woman and a Sperm Is Not a Man

When most humans talk about sex, they are not really referring to biology. More often, they are talking about gender. Gender is the social and structural experiences framed by societies and cultures that includes norms and expectations for behavior and appearance associated with aspects of sex biology.[8] Gender is femininity, masculinity, and other frames for social roles. As such, the terms "man" and "woman" are primarily about gender, not biology. Simone de Beauvoir lays this out in her discussion of "women" in the patriarchal society of her time stating, "One is not born, but rather becomes, a woman. No biological, psychic, or economic destiny defines the figure that the human female takes on in society; it is civilization as a whole that elaborates this intermediary product between the male and the eunuch that is called feminine."[9]

Gender is a crucial factor in human lives because it shapes and constrains behavior, with groups and institutions rewarding/recognizing those who meet gender expectations and often punishing/dismissing those who violate or contravene them.[10] Gender is all around us.[11] Gender roles are a powerful influence when thinking about sex biology because they affect every aspect of daily life for individuals. Gender can shape biology itself. Case in point: a recent study of over a thousand brain scans from twenty-nine countries demonstrated that greater gender inequality in a society seems to affect the neurobiological responses of women differently than in the contexts of societies with reduced inequality.[12]

Gender norms consciously and subconsciously assert how we believe certain bodies and body parts "should" be and how the humans with those bodily presentations "should" behave: for example, who can and should wear a dress, a sari, a miniskirt, tight shorts, or a kilt, and what reactions the "wrong" bodies wearing them elicit. At the same time, society's gender expectations dictate who holds a door open for whom, how one sneezes or belches, and even how one sits or stands. Nearly every aspect of daily life is influenced by the gender-space/role one occupies. Look at upper classes in certain European nations across the fifteenth to eighteenth centuries, for instance. Men wore makeup, long wigs, high-heeled shoes, and tights to convey masculinity. Today, in many of those same regions, these items are associated with women and femininity. Or consider the reality that cultural roles for and social acceptance of transgendered individuals was a typical aspect of many societies' gender systems historically.[13] Such traditions became repressed or banned with the colonial expansion of Christianity and particular political systems over the past five centuries but are now becoming more common in some contemporary cultures again.[14]

Gender, as a cultural experience, is not consistent or uniform. Overviews of global gender dynamics demonstrate substantial cross-cultural variation in how gender plays out. Using a simple woman/man dichotomy or feminine/masculine spectrum does not cover the entirety of gender roles across human societies.[15] This is not to say there are not core commonalities in contemporary gender systems, especially in the context of childcare, labor, economic and religious roles, kinship, political structures, and expressed sexuality.[16]

However, there is rarely a perfect one-to-one correlation between patterns of biological variation and peoples' lived experiences of their bodies through gender. Being classified as, or feeling like, a "boy" or a "girl" or "masculine" or "feminine" does not necessarily correlate to what one's genitals look like, what genetic sequences reside on their twenty-third chromosome, or whether or not one can make ova or sperm. For example, a majority of, but not all, humans with a penis and testes associate with the gendered categories of man or masculinity. But do all those individuals who identify as such have the same bodies? Do they have the same desires and attractions in sexual partners? Of course not. Even in a cluster of individuals identifying as man and masculine there are substantive variations in every one of the key bodily and psychological aspects of that gendered role. Such variation does not make individuals better or worse at the gender role but demonstrates that such roles should not be thought of as uniform, or in binaries, because that is not how people experience them. Also, for many humans, aspects of their sex biology do not "match" their assigned or personally identified gendered categories, and there are those who sense themselves as not corresponding to any of the available categories at all. Gender and biology have a complex relationship—not a 1:1 correlation.

The assumption that gendered differences are primarily or "best" explained by specific differences in sex biology is both oversimplified and incorrect.[17] To assume that everyone who has a vagina, clitoris, and labia, or a penis and scrotum is a uniform kind of human relative to the "other" kind is insufficient and ill-informed. One's sex biology is important and plays a major role in how any given individual engages with gender and other cultural dynamics of their society. But it is neither simple, nor uniform, nor binary. It's biocultural.

Two Terms We'll Use

Given all that we're covering in this book, we need to set some terms to assist with clarity and precision in the remaining chapters. Any effective examination of the human experience in relation to sex biology is always contingent on two things: the sex-biology variation in play, and the specifics of the culturally structured gendered system. It's the dynamic interweaving of these two traits that distinguishes humans from many other animals. For this reason, the best contemporary scientific approach to this discussion is to frame it in terms of "gender/sex."

Gender/sex is the notation used to reflect the intertwined biocultural reality of bodies and experiences more accurately.[18] Think of it as an extension of the ongoing rejection of the nature vs. nurture frame in the study of humans. By adopting a biocultural framing, research on human bodies and lives is no longer forced to arbitrarily, and often erroneously, separate the material form from the lived experience (nature from nurture, or—better put—biology from culture). This has proven extremely important in health research, cognition, neuroscience, and a myriad of other arenas.[19] Today, gender/sex is the term increasingly used by researchers who study sex, biology, gender, sexuality, and behavior in humans to acknowledge the reality that the biological and the cultural are always entangled and often inseparable. So for the rest of the book, we'll use the terms "gender/sex" and "biocultural" in many locations both for clarity as to what we are talking about and to reflect the contemporary state of science regarding humans.[20]

One final set of terms is necessary. Most research on human sex biology uses "sex at birth" as the key category for dividing up subjects by "sex" in the study. However, "sex at birth" usually means a designation based on a classification of the genitals of the newborn. There are usually no tests for chromosomal complements, gonad types, or anything else along those lines. As such, the "sex at birth" category is a little sloppy, and sometimes misses the mark. What "sex at birth" is hoping to measure is really what is called the 3G category of sex.[21] The three Gs are genes, gonads, and genitals. A "3G female" is a human who has XX twenty-third chromosomes, ovaries, and a clitoris/vagina/labia. A "3G male" is a human who is XY and has testes and a penis/scrotum. The assumption is that a look at the genitals gives you a solid inference of the other 2 Gs and thus that this works as the classification system and the working definition of "sex."[22] The rationale for using this categorization is that the 3 Gs are highly but not absolutely correlated with one another. Thus, while 3G categories reflect a set of typically patterned variation related to reproductive biology, they are not even close to 100 percent accurate at capturing the actual range of human biological variation, even in relation to the 3 Gs.[23]

3G categorization *does not* reflect the biological reality for at least 1 percent of all humans,[24] if not more. This represents at least ~80,000,000 humans living right now. 3G female and 3G male are not absolutes, nor are they the only definition of "sex" biologically or of the terms "male" and "female." 3G categories do not always or necessarily map to the terms "man" and "woman," nor are they necessarily correlated with the gender dynamics of masculine and feminine in only one direction or manner. But they are biologically definable categories (if at times incomplete), which is better than "sex at birth," or "man" or "woman," and unfortunately most of the data we'll be working with in the following chapters either assumedly or explicitly uses 3G categories. So when discussing these data, we will be using these categories too.

Now, with a solid grasp of basic animal sex biology, the biocultural reality of being human, and the framing of human gender/sex, we are ready to tackle sex biology and the human experience past and present.

4

Humans Then

IF YOU LINE up a monkey, an ape, and a human, you'll see a lot that looks similar, and some key differences. Then, if you watch their behavior, there'll be some similarities, but many more differences. This is called the comparative approach: comparing one kind of organism with its closest relatives to see what in their bodies and behavior likely comes from a shared past. We can also study an organism's past by looking at the fossil record, the history of its ancestors. Both approaches help us understand human sex biology. Humans are primates, a particular type of primate called hominoids (apes), and a very specific lineage of hominoids called hominins (figure 1). The bodies we walk around in and the societies we live in are shaped by our primate, hominoid, and hominin evolutionary histories.

Each node of the human past (primate, hominoid, hominin) offers evolutionary clues to current human sex biology and how it relates to bodies and minds.

Primates Are Us

Primates are mammals[1] and thus have a reproductive physiology that involves shaping large-gamete-producing bodies for potential gestation and lactation. Being mammals also indicates

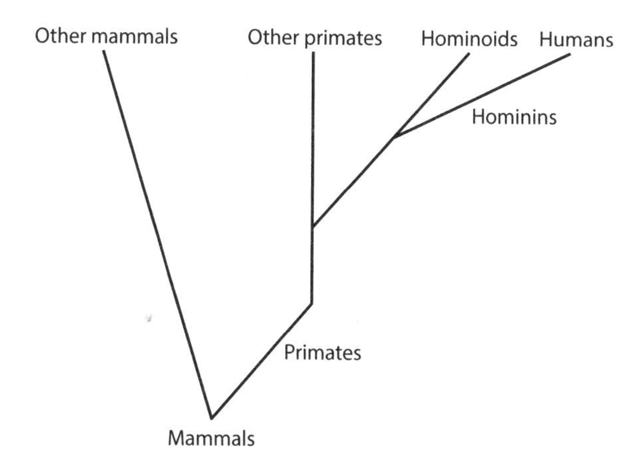

FIGURE 1. The human evolutionary tree.

the presence of an extended system of care of the young after birth.[2] Primates have large, complicated brains and lives centered around social relationships. A key part of these social relations involves very needy infants who are often the center of attention.[3] In many primate species, the mother is the main caretaker, but from shortly after birth infants are moved from arm to arm of other group members, spending time being held by, cared for, and playing with siblings, kin and other members of the group.[4] Some primate species have fathers as the primary or only caretakers, and in others there are no single primary roles, and different aspects of care are spread across multiple group members. Primates as a lineage evolved a diversity of modes of infant care and social life and a varied physiology that relates to both.

Because primates evolved a focus on social relationships as their main way to make it in the world, sexual interaction outside of reproduction is a big part of many species' lives.[5] Primate sex biology and behavior are not just about to getting gametes together. Indeed, many primate species have socially learned greetings, vocal calls, tool use, diverse ways of grooming and developing and maintaining friendships in both sexual and nonsexual contexts. These "cultural"[6] processes play important roles in primates' daily lives. Primates' social behavior and relationships play central structuring roles in how sex biology is shaped and experienced.

Before we explore primates further, a quick note: in keeping with the typical terminology in primate studies, in the rest of this section we'll use the term "male" for small-gamete producers and "female" for large-gamete producers. This shorthand is meant to reflect the biology associated with 3G females and 3G males, not anything about specific behavioral roles or patterns. There is little known about the physiology of intersex nonhuman primates, except that the frequency of such variation is probably similar to that in humans. However, there are few, if any, studies that have looked specifically at variation in 3G categories in primates, but we do know that there is substantial biological and behavioral variation within the categories of "male" and "female" across and within many primate species.

There are important differences across primate species related to sex biology. In roughly a third to a half of primate species, males are a little larger bodied than females (10–20 percent). In a smaller percentage of species, there is even greater variation in body size, with male bodies 25 percent to 100 percent larger than females. In the species where males are substantially larger, they can have huge canine teeth.[7] In such species, males often use their body size as a social tool to dominate and control the group (but not always successfully[8]) and/or to fight with one another over resources.[9] This extreme variation in body and canine size in some primate males is argued to emerge from sexual selection. As Darwin originally proposed, sexual selection results from competition between or within large- and small-gamete producers and is related to achieving successful reproduction. In species with very large-bodied males, females may be shaping males' evolution by preferring to mate with larger males, or males may evolve larger bodies and bigger canine teeth via the pressures of male-male competition (or both).[10]

Usually, sexual selection in primates is seen as being facilitated by the basic differences in sex biology between males and females, with males having low investment in reproduction (beyond copulation) and females having very high investment. Sound familiar? As we've already discussed, this assumption based on gamete size and related investment is not wholly supported, and the original model of what anisogamy implies is not universally accurate. There are many permutations and much diversity in reproductive dynamics across lineages and species. But there are some patterns in primates. In most species with large size variation, females (mothers and others) do the majority of infant care, and males do fight among themselves (sometimes a lot). But there are exceptions. In gorillas, where males can be more than twice the size of females, adult males are very tolerant of young and participate in their care.[11] In macaque and baboon monkeys, where males are up to 25–50 percent larger than females, males typically don't do much caretaking, but it is not totally uncommon for males to adopt and care for infants whose mothers died, and in many of these species, some males spend as much if not more time with infants and young than they do with other males.[12] The bottom line is that in highly sexually dimorphic species of primates, reproductive physiologies usually correlate with some patterned differences in bodies and behavior, and that sexual selection likely had a role in structuring some of these patterns of sex biology.

But the same is not true for species with little or no body-size differences. In these less dimorphic primate species, behavioral differences are small, absent, or, in some cases, in the reverse direction of what they are in the species where males are much larger. In chapter 2, we reviewed the pair-living primates where there is often care of infants either by all group members or primarily by an adult male. Most primates that live in small or pair-bonded groups are relatively monomorphic (same-sized), and in the ones where there is a lot of male care of young, there are substantial commonalities in hormone activity across adults who care for infants regardless of their sex biology.[13] There are also non-pair-living primates that are monomorphic, and in those species, we don't see the pattern of male dominance, nor its biological correlates, common in primates with large size variation between sexes.[14]

While primates come in all shapes and sizes, from the tiny mouse lemur that can fit into the palm of a human hand to the mountain gorillas who weigh three times what most human adults do, they are relatively consistent in one critical way: primate infants are large, slow growing, and fully dependent on care for a longer period that than most other mammals. For primates, the patterns of care for young in a given species often exert substantive influence on bodies and behavior.

Humans Are Hominoids (Apes)

Because we are evolutionarily a part of the hominoid lineage, that group of primates (the living apes) provides a glimpse of what is shared between us and them, offering insight into what is deeply ancestral to human sex biology and the human experience. Thus, it should not be surprising that apes demonstrate a lot of variation in how sex biology plays out in bodies and behavior, within and between species.

Gibbons are the smallest apes and are humans' most distant ape relatives. Gibbon females and males are mostly the same size with similar-sized canine teeth. However, in a few species, males are slightly bigger, especially in the siamang, and in many gibbon species adult males and females have different-colored fur.[15] The most striking difference between many male and female gibbons are their vocalizations. All gibbon species give long melodious vocalizations, called songs, but generally males and females give different versions, often as complementary duets. Gibbons primarily live in small groups, often pair-bonded, with just two adults and some young. Female and male behavior overlaps extensively except that females do most care of the young, with the exception of siamangs, where males who do a bit more care of the young than other gibbon males.[16]

Orangutans shared a common ancestor with the African apes and humans more than ten million years ago. Orangutans are highly sexually dimorphic with adult males up to twice the size of adult females. But there are two different types of "male" orangutan, both basically 3G, but at least one (or two) of the Gs acts differently, and other aspects of their bodies and behavior also differ from one another. Most adult males develop a set of distinctive fatty cheek pads called "flanges" on their face and a large throat sack that enables them to give deep and powerful long calls. However, some adult males don't develop the flanges or do long calls. This "two kinds of 3G male" is an orangutan-distinctive sex biology.[17] The males that do not develop the flanges and other aspects of large adult male appearance seem to not have the circulating levels of a set of hormones (LH, testosterone, and DHT) necessary for development of the flanges and a full throat sack. However, these unflanged males do have sufficient hormone levels and patterns (specifically testicular steroids and FSH) to develop full sexual function and sperm production.[18] Their gonadal and other endocrine (hormone) function is not atypical, but it is different from that of the flanged males. So, unlike most primates, orangutans have three body types related to sex biology: small females, flanged males, and unflanged males, and these types behave differently. Flanged males spend the majority of their adult lives alone, coming together with females for a few days now and then to mate. Unflanged males also spend most of their time alone, but sometimes hang out for a while with another unflanged male, especially when they are young. Females spend the majority of their time with their young offspring and on occasion with one or a few other females. Orangutan social lives and sex biology are distinctively different from those of the other great apes.[19]

Gorillas, as noted above, have enormous size variation. Males can be more than twice the size of females. Gorillas often live in groups with one adult male and a number of females and young. But as many as 40 percent of gorilla groups have more than one adult male. It is not uncommon for many gorilla groups to come together in large multigroup assemblages from time to time with many interactions, including homosexual sexual behavior, between the members of the different groups.[20] While gorilla mothers are the primary caretakers of infants, gorilla males are highly tolerant and offer both protection of the young and often spend a good amount of time playing and interacting with them.[21] Gorillas do map to the general pattern of male dominance in largely dimorphic primates, but they differ from many such species in the fact that males are attentive to, and engage with, the young regularly.

Chimpanzees and bonobos, both members of the genus *Pan*, are humans' closest primate relatives. These two different ape species illustrate a range of behavioral diversity but share a more or less identical basic sex biology. Chimpanzees (*Pan troglodytes*) have slightly more sex-related variation in body size than humans and have much larger canine teeth. They live in communities that break up into smaller subgroups for much of the time, occasionally coming together as the full group. Adult males in most chimpanzee communities spend the majority of time with other adult males, and females mostly spend time with their young and a few other females. Males are often socially dominant to females and use aggression regularly. Sexual interactions occur heterosexually and homosexually as part of the complex social lives of chimpanzees. Mothers do most of the infant care, but older siblings, still with the mother, often chip in. The patterns of relationships across the range of chimpanzee populations (and subspecies) vary. For example, the Western chimpanzee subspecies has smaller groups, less effective male aggression toward females, and more participation in group aggression and more hunting by females (in fact, they are the main hunters who use wooden sticks as spears).[22]

Bonobos (*Pan paniscus*) have approximately the same level of dimorphism as chimpanzees and pretty much the same sex biology, but their social system and behavioral profiles are different. In bonobos, females tend to be the dominant individuals, and whole communities are more often together than in chimpanzees. Unlike chimpanzees, where different communities avoid one another or get into potentially lethal fights, bonobo communities often get together in a calm and friendly social manner (including having lots of sex). While infant care is still largely by the mother, interactions with the young are more widespread and undertaken by a broad range of members in the bonobo community. Sexual interactions are a central facet of bonobo daily life, and nonreproductive sexual behavior is common across all ages and sexes, homosexually and heterosexually.[23]

The striking thing about chimpanzees and bonobos is that they have pretty different behavior and social systems, and patterns of sexual behavior, despite having the same sex biology. And this is not only between the two species: different populations and groups of chimpanzees can also vary quite a bit in their social relations and sex-related patterns. And both species of *Pan* are quite different behaviorally from gorillas and orangutans. The diversity of behaviors and bodies in the hominoids, who all have a very similar sex biology, demonstrates that for these closest human relatives a given pattern of sex biology does not specify one particular set of behavioral and social lives. Sex biology in the ape experience is variable and is not universally tied to one kind of relationship between bodies, behavior, what type of gametes are produced, or which Gs one has: this is the key part of our inheritance from our history as hominoids.

Evolution is as much about discontinuity as it is about continuity. Understanding that humans and other hominoids share an evolutionary history, a continuity, is to acknowledge that this shared past shapes humans' bodies and lives. But that's only part of the story. It's the history of our more recent evolution since our split with the other apes, a discontinuity, where we find deeper and more specific insights into human sex biology and experience today.[24]

Humans as Hominins

The hominins separated from other hominoids sometime between seven and ten million years ago and diversified into a cluster of lineages of ape-like beings who walked upright on two legs. Because we are talking about organisms that lived in the deep past (and all we have are fossils), we can't truly assess any of their Gs. When talking about fossils, we use the terms "male" and "female" as proxies for 3G categories based on typical skeletal patterns for 3G humans today. Admittedly, these classifications are wrong at least some of the time, even in those with bodies just like ours. Remember the "revered leader" case from the opening of chapter 3.

The earliest hominins are not very well known, but by about four million years ago, the hominin genus called *Australopithecus* emerged, and in its early members are the likely roots of the human lineage. The Australopithecines were a bit smaller than most humans today and likely had a good deal of body-size dimorphism, with males larger than females. However, unlike many other size-dimorphic primates, they did not have significant canine-teeth dimorphism, and it's unclear if the social behavior and social systems of the early Australopithecines were more ape-, other-primate-like, or more human-like.[25] The Australopithecines lived in small groups with multiple adults and young, and likely had very strong social bonds between group members.[26]

Between three and four million years ago, the antecedents of the human niche (the way we humans live in the world) were clearly developing. By at least 3.3 million years ago, Australopithecines were making and using stone tools and beginning to interact with, and modify, the world around them in what appear to be precursors to human-like lives.[27] At least some of the Australopithecines evolved a human-like reproductive pattern of giving birth to newborns that need much care and attention from multiple group members in addition to the mother.[28] While this mix of very needy infants with little motor development and lots of neurobiological growth outside the womb is found in other primates and some other mammals, the human pattern of cooperative care of extremely helpless infants is rare and creates a distinctive set of biological and behavioral patterns for human bodies that have much to do with the emergence human gender/sex (see below). Between two and three million years ago, the first members of the genus *Homo* (humans) showed up and began the specifically human story.

Humans as Genus *Homo*

As noted earlier, human babies are born with about 40 percent of their total brain growth completed. Having more than half of brain development occurring outside the womb in the world is a feature not found in any other mammal or primate. Because of the human infant's slowly developing neuroanatomy, coordination, movement capacities, and information-processing skills, they are almost completely dependent on others for the first few years of life.[29] The closest comparison would be a chimpanzee, whose brain is about 60 percent grown when born, but who can cling to her mother's fur and climb around on the mother's body almost immediately after birth. Humans have no fur, and our infants cannot coordinate their limbs or even hold their head up on their own when born. The ability to simply move themselves along the ground, let alone climb on a caretaker, doesn't emerge for months. Human infants must be carried, tended to, protected, fed, and nurtured more than those of any other species. But this helplessness is a necessary step to developing the physiological, psychological, and social pathways to adulthood, and to becoming the neurobiologically, socially, and technologically complex beings we are. The physical and perceptual development of touch, sounds, smells, tastes, looks, likes, movement, and everything else is entangled in, and shaped by, the infant's cultural landscape and all the actors in it. To develop effectively, human brains and bodies, and thus minds, are completely soaked in their cultural environment. This is where the human biocultural experience begins; this pattern evolved in our ancestors.

The high post-birth-brain-growth and bioculture-acquiring infant pattern began in the genus *Homo* at least a million years ago. But one cannot evolve giving birth to helpless infants first and then figure out how to respond to it. The infants would die. For such a system of infant development to evolve, there must have been a serious capacity for cooperative collaboration and contribution to infant well-being already in place. But how did such a caretaking system evolve, and what does that story that tell us about our sex biology?

From its earliest appearance more than two million years ago, the genus *Homo* shows little sexual dimorphism relative to earlier hominins and to most other apes. *Homo* has small body-size variation related to sex biology (~10–15 percent) and no differences in canines. Aside from the typical skeletally identifiable correlates of birthing and slight body-size variation, the fossil and paleoarcheological record hold little evidence of physical demarcations, and even less of behavioral and social variation, between 3G categories in the genus *Homo*.[30]

Compared to other hominins (like the Australopithecines), the genus *Homo* underwent specific morphological changes in the pelvic girdle, legs, feet, arms, hands, skulls, teeth, and faces, alongside less easily measurable, but significant, behavioral and cognitive shifts across the last two million years or so.[31] During this time period, much of what we take for granted as "human" emerged. Things like hypercooperation, complex social interactions, and increasingly complex material technologies such as tools and fire.[32] Members of the genus *Homo* worked together, assisting one another day in and day out, expanding the intensity of mutual care and the depth of social bonds and reliance on each other.[33] Via this intensive cooperation and coordination, *Homo* developed more intricate and diverse foraging and hunting patterns involving not only the use of stone, wood, and bone tools, but also increasingly complex collaborative behavior and communication. Around this time, the brains of the genus *Homo* started getting more neurobiologically complex, necessitating a longer growth and development period: *Homo* childhood extended to be the longest developmental process for any primate (or mammal). By at least about eight hundred thousand to a million years ago, our ancestors began to experiment with fire, eventually working together to use it to reshape bone, wood, and stone, and to cook a plethora of different foods. By three hundred to four hundred thousand years ago, there is strong evidence for social connections across different groups and places, the emergence of long-distance exchange networks, and increasingly dynamic intergroup relations. Then, in the past few hundred thousand years, we see the emergence of engravings, carvings, and eventually art. And all this was accompanied by increasingly complex communication and information sharing, eventually resulting in that amazing system called language.[34]

Emerging from this *Homo* evolutionary history are three key patterns that offer insight into how human sex biology and behavior, our experience of gender/sex, evolved. The first is an intense cooperation and coordination around childcare. The second is increased diversity in social and sexual relationships. And the third is the development of categories of culturally mediated perceptions and expectations of and for different members of the group related to aspects of sex biology, what we call gender and gendered roles.

It Takes a Group to Raise a Child

In most primate species, mothers do the majority of infant care, but some amount of allocare (care of young by individuals other than the mother) is common across the primates.[35] Usually, the allocare is done by females related to the mother, and/or siblings, and/or the father, but in those species that exhibit intensive allocare, the entire range of group members assist in the raising of the young. Allocare behavior by other group members alters the evolutionary pressures on mothers and affects both group members' and mothers' bodies. Hormone dynamics and physiologies shift in the caretakers of heavy allocare species, making them somewhat different from other species with the same basic reproductive anatomy but mother-only caretaking systems.[36] Humans are the only primates that combine intensive allocare behavior with superextended childhood: we have a deeply cooperative and collaborative multicaretaker childcare system, *and* human infants are born in more need of more intense and long-term care and assistance than any other primate. Human infant caretaking evolved as a group effort that reshaped human bodies.

Recent overviews of the available evidence for the evolution of the distinctive human infant caretaking system demonstrate that humans not only evolved a suite of biological processes to deal with time- and energy-intensive infants, but that we also take care of these infants through a variety of cultural adaptations, including material culture, provisioning of food, and shared child care. Humans evolved to be biocultural cooperative caretakers in every sense of the word.[37]

By between a million and five hundred thousand years ago, the genus *Homo* displayed remarkable levels of cooperation and collaboration, more so than any other primate or hominin, and they used this capacity to develop a particular type of caretaking.[38] Caring for infants shifted from a primary reliance on the mother to a system wherein humans were raised by "Mothers and Others."[39] The human infant is a distinctive evolutionary twist on primate babies in that the level of care, attention, and engagement that it requires to develop could only emerge in a system where multiple potential caretakers are already present. The way our lineage enabled such helpless infants to emerge, paving the way to our incredible cognitive and neurobiological biocultural evolution, was the co-option of an already strongly developed pattern of cooperation and applying that to childrearing. Multiple researchers over the past few decades have demonstrated via fossil, archeological, and comparative studies on contemporary human bodies, that caretaking in the genus *Homo* evolved to involve more than the mother; siblings and other members of the group were crucial to the effective raising of a child. Carrying of *Homo* infants, a large cost, was spread across group members.[40] Older individuals and siblings were and are key caretakers, and caretaking by a range of older individuals might even have resulted in the evolution of longer human life-spans.[41] All human bodies (not just 3G-female ones) evolved strong physiological/hormonal responses for caretaking behavior. For example, unlike many mammals, human adult 3G-male bodies change physiologically in the presence of infants, most strongly when they are engaged in caretaking of them.[42] The extent and structure of cooperative parenting and broadscale caretaking in the genus *Homo* created a distinctive context for human evolution and shaped human sex biology.[43]

The genus *Homo* evolved a system where the mother is never alone, so the pressures of successful reproduction are not solely on mothers' bodies. The evolutionary implication of this deeply cooperative and complex caretaking is a rejection of a simple calculus of extreme evolutionary differences for bodies and behavior based on gamete producing and individual costs of gestation and lactation.[44] The data from the fossil record and archeological evidence related to reproduction and energetic investments in the genus *Homo* indicate that we cannot model an understanding of human sex biology simply by separating small- and large-gamete producers assuming a fixed and massive difference of investment in reproductive effort. Nor can we lean solely on the expected constraints of having a uterus and ovaries, or testes, to understand human sex biology's relation to the rest of our bodies and behavior. Indeed, a decade ago, a cluster of biologists[45] specifically reviewing Bateman's principles (the asserted universal impacts of anisogamy) and contemporary sexual selection theory demonstrated that "human mating strategies are unlikely to conform to a single universal pattern."[46]

There are a few other aspects of human mating that differ from those of many other mammals: almost every adult produces offspring, and very few produce way more than others (humans have a very low "reproductive skew"). Humans have the lowest variation between individuals in number of offspring produced and the strongest similarity between females and males in number of offspring produced of almost any mammal. In short, humans evolved a distinctive relationship between sex biology, mating, reproduction, and parenting behavior.[47] Therefore, explanations for human bodies and behavior that rest on the assumption of major evolved differences between 3G females and 3G males related to reproduction and parenting are inconsistent with the existing data on human evolution, and human reproduction, and thus offer insufficient explanatory power.[48] The human lineage evolved a particularly intense, social, and communal mode of reproduction and childcare that shapes how our sex biology works and what it means for human bodies and behavior. But sex biology is not just about making and caring for babies.

Sex Is More Than Reproduction

Genital-genital contact, genital manipulation via hands, mouth, and feet, genital-anal contact, and other forms of sexual activity are found outside of reproductive possibilities in many monkeys, and in all apes and humans. In some primates, social sex is as frequent as reproductive sex (if not more so).[49] In such species, sex is a social tool, and sex biology must be understood in the context of both its role in reproduction *and* its use in social relations.[50]

Both members of the ape genus *Pan* (our closest primate relatives) excel at social sex. Bonobos are famous for frequent social sex, and chimpanzees[51] also participate in sexual behavior outside of reproductive possibilities.[52] Bonobos engage in sexual activity in dominance relations, feeding, group cohesion, play, bonding, fighting and reconciling, and pretty much everything else (including reproduction). Being so sexually active has hormonal and other physiological impacts.[53] Chimpanzees also use sexual behavior as a social tool, but not quite as much as bonobos. And in chimpanzees it may be more common between males than females.[54] In gorillas[55] and orangutans,[56] sex outside of reproduction is less common, but it is present, most often in homosexual contexts. And humans outdo all our primate cousins with an enormous range of nonreproductive sexual activity. In fact, one could argue that most sexual activity in our species occurs outside of reproductive possibilities.[57]

The pervasiveness of sexual activity as a social tool in apes and in contemporary humans suggests that such behavior is an ancestral hominoid state for apes and humans. Across hominoid and hominin evolutionary history, sexual interactions are often detached from reproductive function but are attached to other social and physiological functions. So, any explanation for the evolutionary impact, functions, and constraints or benefits of sex biology cannot *only* be explained by its relation to reproduction. Given that humans are much more socially sexual than any related species (except maybe bonobos), the most likely explanation is that across hominin history, and especially in the last two million years of the evolution of the genus *Homo,* the role of socio-sexual behavior has been amplified as a major aspect of human sex biology. Sex biology is not just for making babies, not in the apes, not in the hominins, and most certainly not in humans.[58] But, humans do make babies, have lots of social-sexual relations, and build amazingly strong social bonds. All of these processes—making babies, having sex, and making bonds—are interconnected.

Social and Sexual Pair Bonds Are Complicated

In the animal behavior literature, a pair bond is a special, predictable relationship between two adults, often a reproductive relationship. Humans frequently have pair bonds. This fact has been used to argue that the presence of human pair bonds stems from patterns of difference in 3G-male and 3G-female sex biology. Some researchers suggest that humans evolved heterosexual pair bonds and the nuclear family because of the specifics of our sex biology. The story goes something like this: given basic differences in mammalian reproductive biology and the costs of human infants, humans who give birth and gestate (largely 3G females) are under extreme pressures, so bonding with a high-quality partner (a 3G male in this proposal) improves her chances of successfully having and raising an infant.[59] This same argument assumes that 3G males would prefer to mate with many 3G females and not invest highly in just one, but the costs of the human infant and the competition for access to the best-quality mothers make some degree of pair bonding a good option (especially if the 3G males can keep the option to "cheat" on occasion open). This story is used to explain the "human nature" of the nuclear family, 3G females as primary caregivers, and 3G males' tendency to seeking more sex partners than females. But is it accurate?

In human evolution, the pair bond has been characterized as a "special and predictable relationship between a male and a female that involves tight social connections and a sexual relationship, and usually includes mating and the raising of young."[60] Many scholars have argued that this "pair bond" set-up is monogamous, is the basis of human society, and evolved in our hominin ancestors due to the assumed core costs and structures of mammalian large-gamete- and small-gamete-producers' bodies.[61] For example, some primatologists argue that the heterosexual pair bond (and monogamous mating) precedes and gives rise to the nuclear family structure in human evolution, and that its appearance is a key point in the evolution of human society.[62] But pair bonds are not necessarily linked to a nuclear family structure in humans or in other animals. In fact, pair bonds are not necessarily linked to reproduction at all.[63] Rather, pair bonds are an effective way to enhance and expand the social networks and cooperative possibilities, and they are not always related to reproduction.[64] And, as far as reproduction is concerned, we know that, over approximately the last two million years, our lineage evolved a complex system for raising and caring for young that involves more than one mother and one child (and one sperm producer). We are not a species that evolved a system of reproduction and social groups based on heterosexual pairs.

More than twenty-five years ago, I began reviewing all the available data for pair bonds in primates and humans.[65] I discovered that primates (including humans) are not more monogamous or pair-bonded than other mammals,[66] and that while there are many primate species that live in small groups consisting of a male-female pair plus offspring, only some of those exhibit pair bonds. Work since then demonstrates that pair bonds come in a number of different types and are rarely associated with actual reproductive monogamy.[67]

It turns out that there are two types of pair bond: social and sexual. The social pair bond is a "strong biological and psychological relationship between two individuals that is measurably different in physiological and emotional terms from general friendships or other acquaintance relationships." The sexual pair bond is a "pair bond that has a sexual attraction component such that the members of the sexual pair bond prefer to mate with one another over other mating options."[68] In mammals, including humans, pair bonds combine social relations with a range of biological activity.[69]

Humans have social and sexual pair bonds. Human social pair bonding occurs across the entire range of gender/sex and age categories, with relatives and with unrelated individuals.[70] Humans have sexual pair bonds both heterosexually and homosexually. And human sexual pair bonding is not necessarily related to reproduction. Therefore, the pair bond, sexual and/or social, while critically important in humans, is not something that emerged directly out of our sex biology or something that is constrained by it.

Gender?

Gender did not appear out of thin air. Humans assign cultural meaning to almost everything in the world, and our ancestors began doing that a long time ago.[71] Cultural meaning is assigned not just to the material world around us but also to different aspects of the human experience, like age and life-stage, body type, hair patterns, genitals, demeanor, voice tenor, whether one gives birth or lactates or not, who one has sex with, caring for infants and others, and much more. The social dynamic that we call "gender" today emerged from complex biocultural evolution in the human lineage. Human gender concepts and experiences have been shaped over evolutionary time by the mix of cooperative caretaking, complex social sexuality, increasingly complex social lives, *and* the specifics of sex biology and other aspects of human bodies.

Gender is a set of expectations, perceptions, and behavior that a social group believes about how bodies and behavior *should be* in relation to aspects of sex biology. Obviously, biology plays a role in gender, but the specific history of a society, its ecology, its demographic makeup, and a whole range of other political and economic factors contribute the specific details of what gender-particular roles emerge in any given group. Humans' deep shared evolutionary history and complex dynamic variation in cultural diversity are why we have some common patterns of gender across societies and so much variation within and between them.[72] However, while we can study gender dynamics in humans today and in the very recent past, it is almost impossible to see gender roles in the fossil and archeological record. That leaves us with the very unsatisfactory answer that the capacity for gender clearly evolved, but what gender roles looked like in the past is not at all clear. Interestingly, what little evidence we do see suggests that if there was something like gendered roles earlier in human evolution, they were probably not identical to those of today.[73]

3G-male *Homo* were, on average, about 10–15 percent larger than 3G females, and we can assume that earlier *Homo* 3G males had, on average, slightly higher muscle mass and density and greater upper-body strength (as they do today). We also know that the nutritional and caloric needs of human 3G females with a uterus go up at the very end of pregnancy and during lactation. This increase in energetic costs for female *Homo* likely started around 1–1.5 million years ago and was part of the evolutionary development of cooperative parenting.[74]

These differences in body size and the fact of costly infants are often offered as clues to gender roles in human ancestors. For example, it is common in recreations of human ancestors to envision a man (an assumed 3G male) standing with a medium-sized dead animal slung over his back (the hunter), a woman (an assumed 3G female) sitting or kneeling holding an infant (the caretaker), and maybe another woman tending a fire with a toddler at her side (the meal preparer), and then another man, usually a bit older, sitting making the stone tools (the tool maker).[75] These images are contemporary gendered imaginings, not data from the past. There is almost nothing in the fossil or archeological record that suggests a specific 3G-sex biology or age makeup or gender role for who did what or that there were "jobs" like this in the past.[76] Rather, for most of the history of the genus *Homo*, there is no evidence for specific roles of "tool specialists," "hunt specialists," "meal preparers," or "childcare specialists."[77] The gendered assumptions of "man the toolmaker," "man the hunter," and "women the caregiver" are recent cultural inventions.[78]

In the comparative sense, we know that ape females use and make tools slightly more than males and that young apes learn to use tools primarily by watching their mothers. In-depth analyses of the reconstructed processes of stone-tool creation and use by *Homo* do not distinguish clear or consistent sex-biology differences in capacities or patterns in tool making and use.[79] Analyses of recent, especially agricultural, tool use and manufacture (over approximately the last five to ten thousand years) reveal some gendered role differentiation in use of tools, but these appear to be largely based on variation related to upper-body size/strength and the requisites of certain types of labor.[80] However, in most *Homo* populations over the last million years, but before the last five to ten thousand years or so, there is evidence of more overall upper-body strength in both those assumed to be 3G males and 3G females compared to most contemporary humans of either category.[81] So, using contemporary humans as the basis to understand ancient *Homo* strength and capacity is underplaying the relative muscular/skeletal strength of *both Homo* 3G males and 3G females in the past and overplaying the relevance of sex-biology-related strength dimorphism for a majority of activities in human prehistory.[82]

In some contemporary human societies, men (the gender) do hunt large game more than women (the gender), and in many societies, mothers, grandmothers, and sisters are the main individuals charged with childcare. However, assumptions about such role differentiation and its patterns in the human past, especially in the context of hunting and caretaking, are challenged via a range of archeological, fossil, and recent ethnographic analyses.[83] This is not to say that various gender roles did not exist. There is some evidence that while members of *Homo* groups likely hunted together, they may have differed in their roles in the hunt and in posthunt processing of foods and hides. For example, recent evidence for hunting in the human past demonstrates not only a high likelihood of 3G-female participation in hunting (and other physically demanding activities) but show that they probably excelled at endurance aspects of hunting endeavors.[84] In another example, Neanderthal teeth from three different sites show cultural wear patterns, likely due to processing of meat and hides associated with hunting, but there is some indication of slightly different wear and tooth-chipping in teeth classified as 3G male and 3G female. This suggests that there may have been slightly different "jobs" regarding the use of teeth in these Neanderthals. What these differences could be is hard to determine, but the fact that the patterns are there in the fossil teeth suggests that some form of gendered differences may have existed. But they weren't necessarily the same gender roles we see today.[85] In the case of caregiving, we've already seen the evidence for human cooperative parenting, but at a broader level, there is also robust evidence that care for the injured, sick, and elderly by all group members emerged early on in the genus *Homo*, and that coordinated and extensive compassion was a major element in the success of our genus.[86]

Art is another area of frequent assertions about paleo-gender. There is very little evidence of any gendered patterns in any form of art before the last four to five thousand years, except a recent survey of one kind of cave art: the outline of hands. Archeologists examined thirty-two of these hand images from eight different cave art sites and calculated whether the hands were most likely male or female, based on the assumptions about the human hand's second digit to fourth digit (2D:4D) ratio, and whether they were adult or youth (based on size).[87] The initial conclusion was that about 75 percent of the hand stencils were likely done by females (smaller 2D:4D ratios). A sex difference! Or not. It turns out that while the 2D:4D ratio has been argued to be a good measure of patterned variation between 3G males and 3G females (in mostly UK, European, and US test subjects) and assumed to reflect differences in androgen hormone levels during fetal development (higher in 3G males than in 3G females), it doesn't. 2D:4D variation between 3G males and 3G females varies by population, is sometimes absent, and most likely reflects gendered activities rather than a result of universal sex-biology differences in prenatal androgen exposure.[88] 2D:4D variation does not work as a sex- or gender-classification system across all living humans, or for human ancestors. Interestingly, at least five of the hand prints appeared to be those of teenagers, and in other studies these same kinds of hand stencils often show that many of the artists' fingers were cut at the knuckles or were missing.[89] It is not at all clear if this age pattern and the mutilation of hands tells us anything about gender, but it does show that the making of this kind of art was certainly a process that resulted from the interweaving of biological and cultural dynamics.

Many scholars note that heightened social and material complexity (more complicated technology, political systems, economies) and inequality, all usually associated with patterns of contemporary gender roles, do start showing up in many, but not all, societies, in the most recent phase of human evolution, which is the last five to eight thousand years or so.[90] This suggests that the current sex biology of the genus *Homo,* which has been around for much more than the last eight thousand years, did not create contemporary gender roles.[91]

Clearly, body size, strength, and reproductive physiology all play substantive roles in human society and its structuring of gender roles, but not always in consistent manners nor necessarily in the same manners across time and culture. In the last five to eight thousand years, contemporary gender roles became widespread across our species, initially seen in the archeological evidence of burial patterns and grave goods.[92] But the patterns of inequity and gender roles that emerged and became common in certain places and times do not always map to the majority patterns of the present day—remember the Iberian revered leader from chapter 3 and the many recent archeological finds of ancient "warriors" that were assumed to be 3G male but have turned out to be 3G female, at least in one or more of the Gs.[93] Over the last six thousand years, researchers find increased evidence of difference in the bone and tooth chemistry between 3G-male and 3G-female bodies, suggesting social differences causing differences in nutritional status. Muscle scars and wear marks on bones at archeological sites suggest slightly different lifestyles or work patterns between people, with some of these patterned differences emerging between 3G males and 3G females (as best we can assess 3G sex from the remains). There is also an uptick in the birth rates and a reduction in the time between births in the past five to eight thousand years, suggesting that in many groups 3G females who can give birth were more frequently pregnant and occupied in lactation than at earlier periods in human evolution. These biological markers indicate the presence of behaviors and lifeways that seem to reflect specific culturally shaped social dynamics and perceptions that mark the emergence of some contemporary gender roles.

Humans Then Shaped Humans Now

The cumulative evidence from the human evolutionary past demonstrates that anisogamy and sexual selection hypotheses based on less than solid assumptions about large- and small-gamete-producing biology do not offer adequate insight for understanding the evolution of human bodies or behavior. Human evolutionary history is much more than gamete production, mating competition, and simplistic assumptions about sex biology and gender roles. Humans have evolved diverse modes of family and community and an extremely dynamic and very social sexuality. While recent gendered roles are broadly distributed today, they are just that: recent. This brief journey through the evolution of human bodies and behavior sets the stage for the next necessary set of information: the details of sex biology and its patterns of variation in humans today.

5

Humans Now

PUT TOGETHER a random set of two hundred people and invite fifteen 3G males and fifteen 3G females from that group to step forward. Line them up by height, tallest to shortest. Looking over the lineup, you won't see all 3G males on one side and all 3G females on the other. Instead, there'd most likely be more 3G males in the taller half and more 3G females in the shorter half, but individuals of both categories would be interspersed across the entire lineup. That 3G males are on average 10 to 15 percent larger than 3G females in our species does not mean that *every* male is larger than *every* female. It just means that the averages in height between 3G females and 3G males are separated by that percentage.[1] Height in humans is not a sex binary and not a true dimorphism. Height is one morph (a measurable shape) with a range of variation that *can* be divided into overlapping clusters composed of 3G males and 3G females. But it does not automatically have to be divided that way. Height distribution can also be sorted by age, by people from different latitudes and dietary practices, by athletes versus non-athletes, and by a range of other variables depending on what questions you are asking. There are not two forms of human, a tall and a short version; rather, there is a range of variation with some patterns in that variation.

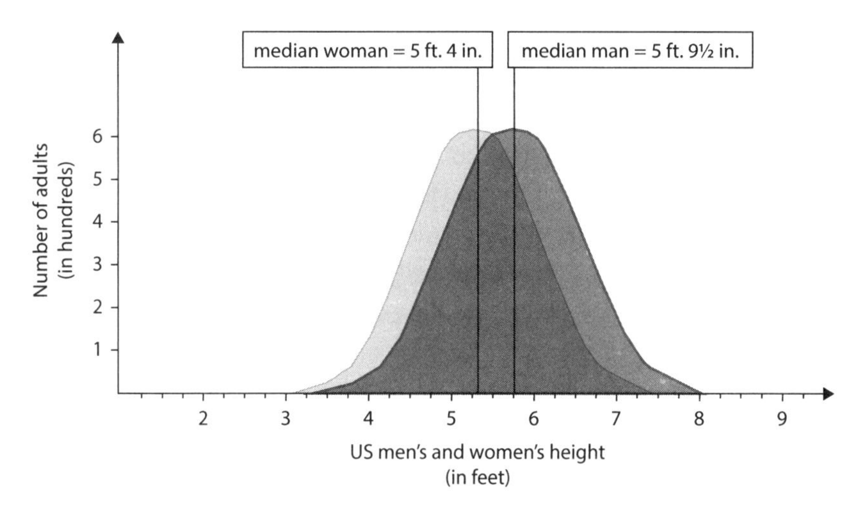

FIGURE 2. US variation in height distribution in "men" and "women." From A. Fausto-Sterling. 2014. "The Science of Difference." *HuffPost,* May 23, updated December 6, 2017. https://www.huffpost.com/entry/the-science-of-difference-lets-do-it-right_b_5372859. Data from the CDC: https://www.cdc.gov/nchs/fastats/body-measurements.htm.

Take, for example, figure 2, which shows the usual graph for the range of heights in the United States when comparing 3G-male and 3G-female bodies.[2] One can focus on the median or on the shaded clusters, but what is most significant, and often overlooked, is that if one takes away the different shading by 3G category and the median markers, one is left with a distribution of height wherein ~78 percent of individuals overlap and are not sortable, by height, into a specific 3G category. So, despite an average difference in height between 3G male and 3G females in the United States, height is an extremely unreliable way to identify any individual's 3G category.

The majority of human sex biology works in the same way as height does: it is morphology with a range of expression across human bodies. This range of variation can often be broken down into overlapping clusters via placing the bodies being measured into the categories of 3G females, 3G males, and people who do not fall neatly into 3G categories. But doing so automatically, before one has a specific question about the aspect of biology being measured, can obscure important patterns and mislead one into thinking that 3G categories are the main categories of key relevance. For example, while it is not apparent in the black and white image, figure 2 (the original) uses gendered (cultural) shading patterns (blue and pink) to identify the clusters of 3G individuals and ignores non-3G individuals completely. Even though the graph is of a measurable piece of human biology (height), it is presented in a heavily cultural fashion (as overlapping gender-colored categories). What if we were interested in the effect of childhood diet on height? Or about the length of the femur (our largest leg bone) relative to the overall height? If we break the sample pool into male and female before we take our measurements, we might miss some very important variation that is not related to sex, or we might see the variation as sex-related when it might not be. The immediate breaking of a sample pool into male and female is a pattern of data presentation, analyses, and discussion that sometimes makes effective and accurate conversation about, and understanding of, human variation difficult.

Most work on sex biology is based on specific assumptions about "sex differences" and thus is framed as a female vs. male comparison. Researchers rarely ask the initial question about the overall patterns of variation of the specific trait or variable being measured (height, muscle strength, hormone levels, etc.) for the study subject pool. Instead, the subject pool is usually divided into "male" and "female," without any concrete definition of those terms beyond self-reporting of assigned sex at birth (based on genitals), *before* the data collection starts. The problem is that if one assumes there is a difference and sets up the research project to look for that difference, the likelihood of finding difference is pretty high.[3]

Because of the widespread pattern of assumed difference in "males" and "females," and the assumption that anyone not fitting into those categories can be ignored, the methods (and theory) in the vast majority of research into sex biology are conceived, conducted, and presented as the study of "sex differences" and not as the study of variation in sex biology. In this book, we are interested in the actual variation in and the patterns of sex biology. So, while we review and discuss a range of studies on contemporary human sex biology in this chapter, we are stuck with the ways in which the studies were designed and the terminology, and methods, based on assumptions of difference they used. This makes our task, trying to see what the actual human variation pattern is before assuming that "sex" is the best way to divide up that variation, a little harder. But we have to use the data and analyses that are available. I'll try to point out the issues with certain studies and some of the more egregious biases as we work through the data, but I would suggest keeping these potential biases in the forefront of your mind as you navigate this chapter.

Humans vary in many aspects of our bodies, inside and out. Much of this variation comes in particular patterns. Some of it is directly related to sex biology. But if we focus primarily on averages and see everything from a core assumption of difference, we often obscure the biological reality of the trait in question. When thinking about bodies, we should consider the forest of variation and not focus only on a few average trees, certain tree clusters, or predetermined ways of looking at the forest. This chapter's brief survey of human bodies, hormones, reproductive physiology, brains, and DNA reveals a lot about human variation, sex biology, and the human experience.

Human Bodies Vary a Lot

All humans share the same physical structures. Human tissues (tendons, ligaments, bones, blood, skin, etc.) are all made of the same materials. But tissues and the bodies they compose can form a variety of shapes. In the human species, body mass (the relation of weight to height) varies by as much as 40 to 50 percent across populations, and in sheer weight, the heaviest humans are as much as four times as heavy as the lightest. The width of the human body at the pelvis (the hip) varies by about 25 percent across our species, and average body heights range from about just under five feet to about six feet (from about 150 to 185 cm). Over the last 20,000 years, our species, on average, has decreased in overall body mass, but recently in many areas of the planet, as health care and nutritional patterns have shifted, people have shown dramatic increases in height, adipose accumulation, and body mass.[4] Some of this bodily variation patterns within and between 3G categories.[5]

The USA National Health and Nutrition Examination Survey (NHANES) dataset illustrates some of this variation. NHANES is a collection of physical and social data drawn from a nationally representative sample of about 5,000 persons each year since the 1960s, and individuals are identified by 3G category before data are collected.[6] In the 2015–2018 USA NHANES[7] dataset, non-pregnant 3G-female adults over 20 years old weighed between 110 and 263 lbs. and 3G males over 20 between 136 and 287 lbs. In this same dataset, the 3G females range from 4′11″ to 5′8″ and 3G males 5′4″ to 6′2″: so, there's clear differentiation at the extremes, but the distributions overlap extensively. If one takes off the 3G labels and just describes the variation, the majority of individuals fall into the space between 136 and 263 lbs. regardless of 3G category, but there is more differentiation in 3G categories at the low and high extremes. These data represent the 5th though the 95th percentiles, so 10 percent of all individuals in the dataset group fell above and below the core numbers in height and weight, adding a bit more complexity—some individuals in both 3G categories are even bigger and smaller than these numbers reveal. Keep in mind that the weight data could be as much a 50 percent different and the height data as much as 15 to 20 percent different if measured in other parts of the world. Human bodies vary a lot.

There are also patterns of variation in skeletons. While all humans have the same bones in pretty much the same shapes (with the size variation as noted above), there are a few skeletal elements that vary, on average, across and within 3G categories. The shape of the pelvic girdle is, on average, more "flared" in 3G females with a larger central space than in 3G males. The common argument for this difference is that a large-headed human infant requires a bigger pelvic opening in females than males. But recent work challenges this explanation, demonstrating that larger pelvic openings in females are common in many primates and some other mammals with smaller infant heads. Pelvic girdle variation is not just about birth requirements. A wide array of patterns in organ size and positioning, plus hormone and developmental processes, influence adult pelvic shape and size, and there is a very large degree of overlap between 3G categories.[8] Remember the "revered leader" skeleton? One of the reason scholars labeled the skeleton a "male" was due to the assessment of the "male-like" pelvic girdle. And they were wrong. The pelvic girdle fell into a range where there is a little less overlap between 3G categories, but there was overlap nonetheless. Ignoring the possibilities of variation in favor of an assumption of fixed difference can lead to problems when assessing skeletons, an issue front and center in the forensics world.[9]

In addition to the patterned pelvic variation, there are some patterns in skulls. 3G females tend to have more vertical foreheads, smaller browridges above the eyes, fewer bony buildups around muscle attachments, and smaller mastoids (the bony protrusion just behind and below the ears). Also, the angle of the mandibular ramus (the back of the jawbone) tends to be larger in 3G females. These patterns are common, but there is huge overlap between individuals in the two 3G categories in every one of these features. For example, I am a 3G male and my mandibular ramus angle is large. Also, the relative appearance and differences in many of these areas of the skulls are affected by muscle action during life. So, the mastoids and neck muscle attachments of 3G females who spend their lives carrying weight on their heads will be larger than those who did not grow up with this practice (and possibly larger than those of 3G males as well). The shape of skeletons differs across many variables, including some patterns between 3G categories, but there are not two distinct sex morphs of the human body.

Muscles Shape Bodies

As muscles act, they pull at the surface of the bones, causing alterations to those bones, affecting the shape of the body. 3G males, on average, have greater muscle mass than 3G females of the same body size, and on average muscle strength in the upper body is greater in 3G males than in 3G females. These variations in musculature, combined with variation in height, are a major reason why, on average, bodies can look different across 3G categories.

A recent review of strength training, compiling data from over five hundred published sources, helps contextualize some of this variation.[10] Muscle strength is directly related to muscle architecture, the neural drive to the muscle (called voluntary activation), and muscle fiber composition. The patterns of variation in muscle strength appear to be explained by both overall body-size dimensions (larger bodies have more muscle) and differences in muscle mass. But training, lifetime activity, and age also affect muscle performance and mass. 3G males on average have more muscle mass than 3G females in absolute terms (as a proportion of total body mass) and carry a greater proportion of muscle mass in the upper body. This pattern emerges around fifteen years of age, and by adulthood patterns of differentiation in strength, on average, between 3G categories are more pronounced in upper-body than lower-body muscles.[11] When measured via various resistance, strength training, and controlled exertion methods (pushing, pulling, and lifting weights), adult 3G-female upper-body muscle and trunk strength averages around 60 percent of 3G-male upper-body strength. On the same measures, 3G-female lower-body muscle strength is usually about 70 percent of 3G-male lower-body strength. These figures all come from studies where the 3G categories are pre-identified and the tests are explicitly comparing those categories (usually not also looking at within-category variation or patterns of overlap of the variation). These data are also composites of many studies where overall size differences between bodies and differences in muscle mass and density are not always controlled for. So, it is not totally clear what role size difference as a category is playing relative to 3G sex as category of comparison. This is relevant as actual strength measures of what different muscles do in real life (as opposed to in the gym or lab lifting weights) are often smaller in between category comparisons with more overlap. But some patterns of between 3G category measured strength still exist, to some degree, when comparisons are made between same height and weight 3G-female and 3G-male individuals. This may be due to on average variation in patterns of muscle fiber in 3G males and 3G females, but it is not clear if those differences are due to differential gendered experiences and/or training, developmental sex-biology differences, or a combination of the two.[12]

Much research controlling for the same muscle thickness (and matched muscle size) between 3G males and 3G females demonstrates some level of greater strength performance in males in 76 to 88 percent of the comparisons (so 12 to 24 percent of the comparisons showed no such difference between 3G categories). But other research comparing muscle strength in similar manners does not reveal the same patterns.[13] In power-lifting competitions, 3G males can usually outperform even heavier-bodied 3G females.[14] This suggests that for at least some muscles (mostly in the upper body), this variation in strength is not due simply to size but to the actual structure and functioning of the muscles. However, we also know that actual structure and functioning of the muscles is not a reflection exclusively of one's biology, because gendered experiences profoundly affect that biology.

In all these assessments of muscle strength comparing 3G males and 3G females, there is significant overlap. Unfortunately, the vast majority of such muscle-strength comparisons do not control for lifetime activity or kinds of training and different effects of societal gender norms and expectations for activity patterns of bodies in the different 3G categories: for example, what kind of play/exercise boys and girls are expected, or allowed, to undertake and how these affect muscle and strength development in growing bodies. While all humans can increase muscle size and strength with strength training, increases in upper-body, but not lower-body, strength are more substantial for young adult 3G females than for young adult 3G males (maybe due to the starting-point differences), and older adult 3G females increase relative lower-body strength more than older adult 3G males via strength training. Finally, a preponderance of evidence suggests that patterns of variation in strength between 3G-male and 3G-female bodies are *not* due to differences in voluntary activation (the brain-muscle connection), but rather to on average variation in muscle mass, size, and density. That is, there does not appear to be any sex-based difference in the neural functioning of the muscular system, so it remains remain unclear if there is a set of different gene activity related to muscle strength differences, or if the differences are primarily due to muscle structure and development.[15]

Previous and ongoing physical activity, training, and lifetime behavior make a major difference in adult muscle performance.[16] For example, the speed of the ball in the fastest-kicked goal in the women's World Cup in 2023 was greater than that of the fastest goal kicked in the men's 2022–23 Premier League season.[17] This is likely due to the increased quality in training and a lifetime of sports engagement by today's women's football (soccer) players. So some, but not all, of the variation in 3G sexes' muscle performance and the overlap between them may be reduced, or increased, via gendered norms and behaviors favoring or restraining physical exercise/activity. For example, many elite female athletes are often coached and trained in a way that minimizes, rather than maximizes, bodily bulking up, as the importance of appearing feminine is weighed against performance goals, both because of broad cultural norms and because of the importance of looking feminine to get endorsement deals. There are also culturally mediated gendered differences in body image dissatisfaction across genders in a range of sports and athletic practices that also impact or restrict how different individuals train and shape their bodies.[18]

In much of the contemporary world, there is an increased focus on, and participation in, activities related to muscle development and training in boys and men (who are most often 3G males) relative to girls and women (who are most often 3G females). Across different cultural gendered landscapes, one sees a range of different demands on the musculo-skeletal systems in development from infancy to adulthood, in gender/sex dynamics, and with age. Little of this cross-cultural variation has been measured, and few studies focus on non-athlete or non-trained individuals. Most humans are neither athletes nor strength trainers. Also, while the relative measures of muscles mass and strength patterns represent a large number of samples, they are mostly from specific cultural exercise and training contexts and in WEIRD[19] nations (primarily European Union, United Kingdom, Australia, United States, and Canada), so there may be some important differences in such patterns across a more global comparison of human bodies.

While some overlapping but patterned variation appears in body composition and in muscles, it's in the genitals and reproductive tract that well-known sex-biology variation, and lesser-known sex-biology similarities, are found.

Genitals, Gonads, and Uteri, Oh My!

Despite the popular use of genitals to classify someone as male or female, genitals don't come in just two distinct forms. In the nineteenth century, it was clear to scientists studying human biology that a range of external genital morphology was possible, and there was no way to sort it into two totally distinct categories. Because genitals did not work as a distinction, by the late 1800s researchers looking for markers of a binary sex biology moved to the gonads, identifying those with testes as male, and those with ovaries as female. However, very quickly thereafter, improvements in microscopy and surgical techniques (especially biopsies) quickly revealed that some individuals had both ovarian and testicular tissue in the same body, or even the same gonad, and some had gonadal tissue that was neither.[20] Both genitals and gonads have particular patterns of variation resulting in a set of typical morphology and function, but they are not binary.

Human genitals emerge after the first six weeks of development from masses of embryonic tissue called the urogenital folds.[21] One element in the urogenital fold, the genital tubercle, develops into either penile or clitoral forms. Another part forms the labia or scrotum. This process is not directly based on what genes one has but rather by the actions of hormone and tissue interactions—although there is a strong but not 100 percent concordance of having certain genetic sequences and the outcomes of this process.[22] Another area, the gonadal ridge, responding to gene activity from multiple chromosomes, creates a suite of hormone-tissue interactions central in gonad formation.[23] The end points of gonad formation are usually testes or ovaries, which become involved in cholesterol-derived steroid hormone production in interaction with hypothalamo-pituitary axis (HPA), a major neuroendocrine system, and gametogenesis (the production of ova or sperm). While it is not common, gonadal development does produce ovotestes and/or both ovaries and testes in the same body as a recurring component in the range of human sex-biology variation.

In most humans who are XX for their twenty-third chromosomes and have no testes, or those who are not XX but who experience early life genetic activation and developmental processes similar to the pattern in those having two X chromosomes,[24] a uterus develops via the fusion of a set of tissues (the Mullerian ducts) around and in the urogenital folds between the mesonephric ducts and the gonadal ridges. The vagina emerges from interactions of the Mullerian ducts, the urogenital folds, the Wolffian ducts, and other tissues, but the exact details are not well understood.[25] This internal set of developments (vagina, uterus) seems facilitated by a range of processes centrally connected to actions of a cluster of gene activation and inactivation combined with estrogens and related hormones between the tenth and twentieth weeks of embryonic development.[26] We'll loop back to uteruses below.

The amazing thing about the genitals and gonads is that, despite what seem like pretty different endpoints in most people, the various versions are all made of the same stuff. Adult external genitals are *usually* observably different and categorizable as such (for example, a penis and scrotum or a clitoris and labia), but there is substantial commonality across a range of shape and size. This commonality is clearly seen in sexual response physiology: in the clitoris and penis, erectile tissues function the same way. Understanding this developmental reality helps us contextualize the large range of variation in human genitals. Penis size and shape vary. Clitoris and labia vary. And it is not always evident whether an infant has a clitoris or penis at birth.[27] Most importantly, most of this variation is irrelevant to adult reproductive function; size and shape don't matter that much at all.

Genitals do not equal "biological sex," or "male" or "female," but their adult forms do reflect critical sex-biology variation and patterns related to reproduction and other physiological processes. And speaking of physiological processes, the very visible and often misunderstood dynamics of fat (adipose tissue) plays a central role in how we both see and understand sex biology in bodies.

Fun Facts about Fat (Adipose Tissues)

Fat matters. It shapes bodies, interacts with hormones, provides energy, regulates physiology, and has cultural meaning.[28] Plus, adipose tissues are part of our sex biology. Humans deposit fat more or less the same way that most mammals do, but we have very chunky infants (a high fat-to-bodyweight percentage at birth) and "wear" our fat a little differently. Biologically, "fat" is adipose tissue and comes in two types, white and brown. White adipose is mostly laid down under the skin in the abdominal, lower back, and gluteofemoral regions (butt and upper legs) and around the organs. Brown fat is mostly deposited around the shoulders and neck. While there are typical human-wide patterns of deposition, specific patterns of where and how fat gets laid down vary by geography, populations, 3G sex, age, familial groups, and individuals.[29] The pattern of fat deposition is the same for 3G sexes until puberty, when 3G females usually begin to increase total fat mass, eventually developing about 10 percent more total fat, on average, than a 3G male of the same height and weight.[30] Deposition location also varies, with adult 3G females depositing more fat around the chest area and gluteofemoral region and 3G males more around the abdomen and the internal organs (usually). These patterns are on average, and there is a huge range of variation from individual to individual, and across the entire species.[31] For example, breasts in humans with testes and no ovaries or uterus are not uncommon. Approximately 30 to 60 percent of 3G males experience a degree of adipose (and glandular tissue, see below) deposition in their chest leading to the developments of breasts at some point during their lifetimes (called either gynecomastia or pseudogynecomastia or both).[32] In humans who menstruate and become pregnant, pregnancy and menopause are both associated with changes in fat mass and its patterns of deposition, but the specifics of how this pattern plays out vary dramatically across geographic and cultural contexts. There is mounting evidence that multiple X and Y chromosome genes (plus many other genes), epigenetic processes, sociocultural stressors, and trauma influence specific adipose deposition and metabolism. And there are even mosaic effects (where a person has two or more genetically different sets of cells in their body) on different adipose locations within the same individual.[33] Adipose deposition, fat, while showing some key average patterns, is quite variable across and within 3G-sex categories in humans.

It's worth noting that human breasts and butts are distinctive largely because of fat and bipedalism. And breasts and butts often play culturally salient roles in our gender/sexed lives. The basic mammalian pattern of adipose deposition, the fact that humans walk upright on two legs, and that we only have two nipples (most mammals have multiple sets), combined with the pull of gravity creates a distinctive look for humans. This "look" involves breasts in most 3G females postpuberty (and in some 3G males) and pronounced buttocks for all humans, with potentially relatively larger buttocks in those with ovaries and uteruses. This gives human bodies a different look than other primates. The more frequent and greater development of breast tissue in 3G females, the slightly greater adipose deposition rates postpuberty, and average body-size differences (smaller) accentuate the appearance of breasts and butts on their bodies. This small pattern of variation in fat deposition frequently forms a basis for perceptions of 3G bodies as being more different than they are. Another visible part of our bodies that plays an outsized role in our interpretation of gender/sex and assumptions about differences is hair.

Human Hair

There are two key types of hair in humans. Vellus hair—nonpigmented, soft, and small—covers much of the body (and is hard to see). Terminal hairs are longer, more rigid, more pigmented, penetrate further into the dermis than vellus hairs, and are what are what most think of as "hair." Growth and thyroid hormones are the main actors shaping hair patterns, but in certain areas of the body (face, genital region, upper torso, thighs) androgens (see "Hormones Are Complicated" below) also play a major role.

Humans are weird when it comes to hair. We have relatively scarce amounts on our body core, yet dense patches on our heads, armpits, and around the genitals. 3G males have, on average, more and denser terminal hair than 3G females, but there is a lot of overlap between individuals within a group and sometimes extreme differences between populations of humans around the planet.[34] There are groups of humans where both 3G males and 3G females have very little body terminal hair and other groups where all members of a group have substantial terminal hair. In some groups, 3G males have higher densities and distribution of terminal hair on the face than in other groups, and it is not uncommon for 3G females in those groups to also have greater facial hair. Hirsutism, defined as the presence of "excess" body or facial terminal hair on 3G females, is measured relative to a "standard expectation" for hair growth. This is a cultural, not a biological, measure. Using this measure, 10 to 15 percent of those who identify as women get labeled hirsute across the human species, with some human groups having ~35 percent of 3G females incorrectly labeled as having "male-like" hair patterns. This is incorrect because "hirsute" patterns of hair growth are well within the typical range of variation for our species, so labeling them "male-like" is a gendered (cultural) classification, not a biological one. There are vastly different levels of terminal hair growth on nonscalp bodily areas across and within different groups of people in our species, and thus many 3G females in some populations will have higher levels of terminal hair than 3G males in other populations and vice versa.[35] We know that terminal hair patterns are connected to androgens, estrogens, growth, and thyroid hormone dynamics and population developmental differences: how much and what type of hair one has does not simply boil down to a "sex difference," as the patterns and variation are related to a variety of physiological, geographic, genetic, developmental, and sex-biology parameters.[36] Human hair does not come in a binary, but rather a spectrum with some hormone-related effects that often, but not always, pattern with respect to 3G categories.

So far, we've focused primarily on the visible differences: bodies, muscles, genitals, fat, hair. But what about our insides?

Organs Are Biocultural

Most human organs don't seem related to sex biology. Kidneys, livers, hearts, pancreases, gall bladders, lungs, and so on are not discernably "male" or "female" except that, on average, the organs in 3G-male bodies are larger than those in 3G-female bodies. Basic organ function does not vary "by sex," and 3G sex is not a determining factor in decisions regarding transplants. While human organs are not biologically "sexed," some are gender/sexed: they are shaped in slightly differing manners as they develop inside the gendered body. For example, on average, 3G-male kidneys are larger, but they also, on average, have more and/or larger nephrons (the part of the kidney that filters your blood) than those in 3G females, which might affect function.

There is also evidence from rodents, and a few human studies, that long-term exposure to different hormonal levels and dynamics (as happens in human bodies via age, 3G sex, and other physiological and life-experience variables) affects the behavior of endocrine receptors on certain organs and potentially some of their physiological characteristics. Higher-circulating testosterone, or more dramatic cycling of progesterone and estradiol, might reshape some of the ways in which a given organ functions. Also, inflammation, which can be heavily impacted by pregnancy, physiological/social stress, and a range of cultural and nutritional dynamics, can also affect organs. A lifetime of being in a particular body in a particular culture can shape organs in specific ways via patterns of endocrine exposures, pregnancy (or not), alcohol and drug use, social, economic, nutritional, and psychological differences in lived experiences of gender/sex, age, race, socioeconomic class, caste, and so on.[37]

Hormones Are Complicated

The endocrine system, our hormones, is implicated in almost everything we do. When considering sex biology, researchers and the public often turn to hormones as a (or the) core differentiation between females and males. But hormones don't create bodies and don't make "males" or "females." Rather, endocrine systems are important but not exclusive components in the development of bodily phenotypes (what bodies look like) and how they function.[38] Hormones via the endocrine system are how the body gets cells, tissues, organs, and other parts to communicate and interact for critical processes, like growth, metabolism, and reproduction.[39] The endocrine system consists of the pituitary gland, the hypothalamus, the thyroid and parathyroid glands, the adrenal glands on the kidneys, parts of the pancreas, the gonads (testes and ovaries) and the hormones that each of those glands/organs produce. Hormones come from, and mediate, genetic action and changes in behavior and physiology in response to varying aspects of physical, social, and developmental environments and experiences. And they do so not by themselves, but via an intricate system of varying hormone receptors and a network of feedback loops within the endocrine system and between the endocrine system and various organs and other physiological components of bodies. Thinking that measuring individual hormone types or levels is sufficient information for understanding how this system works is incorrect and unscientific.[40] Plus, all humans have the same hormones.[41] There are no "male" or "female" hormones. Rather, levels and actions of specific hormones, hormone receptors, and their outcomes vary with age, gonad physiology, genetic variation, life experiences, social contexts, nutritional state, reproductive state, and much, much more. Endocrine system activity is deeply biocultural.[42]

The main hormones secreted by the gonads are estrogens, progestogens, and androgens.[43] Both ovaries and testes produce all three (and a few others), but the patterns of production vary dramatically, with ovaries capable of producing higher levels of estrogens and progestogens, and testes capable of producing higher levels of androgens (specifically testosterone). The gonadal production of hormones is directly connected to and mediated by the hypothalamus and the pituitary. Ovaries and testes are variants on the same theme, not different kinds of organ.[44] However, gonadal actions and their outcomes vary in patterned manners.[45] Usually, there is a concordance between the presence of ovaries or testes and specific other Gs (genetics and genitals). However, there is a nonnegligible range of variation in the specifies of gene action/inaction, gonadal development, and other endocrine and physiological dynamics such that the circulating levels of gonadal hormones, and their effects, across individuals fall outside of what are seen as typical 3G patterns with regularity. Thus, 3G patterns are not always accurate predictors of endocrine status, as there is much variation in the system.[46]

Hormones also vary across the life cycle. Some initial variation in the circulating levels of testosterone and estrogen in bodies emerges in the late embryo stage. Just after birth, the hypothalamic-pituitary-gonadal (HPG) axis is activated, and humans undergo a "mini puberty" until about 3–6 months of age. There are peaks in estradiol and testosterone that act to facilitate the developmental trajectory of the genitals and other reproductive anatomy (glandular development around the nipples and ovarian and testicular gamete-production areas).[47] This early process probably sets the stage for later functioning of estrogen and androgen receptors and aspects of their variation and action in different bodily areas. Differences in fluctuating gonadal hormone levels disappear in most bodies from ~1.5 years of age until between ~8–10 years of age. At about 6–8 years of age, humans start adrenarche, the adrenal glands and the pituitary slowly increase the production of the hormones DHEA, DHEAS, and ACTH, setting the stage for actual puberty.[48] Typically starting between 10 and 14 years of age, puberty includes the full maturation of the HPG axis and the genitals (called gonadarche) and the development of patterned changes in hair growth, muscle mass, fat deposition, breast tissue, an acceleration in height growth, and, in those with ovaries, the occurrence of menarche.[49]

Menarche is the onset of menstruation and the specific ramping up of the hypothalamic-pituitary-ovarian (HPO) axis. From puberty forward, patterned variation in the levels of circulating gonadal hormones becomes common between individuals with ovaries and those with testes. Later in life (about fifty to fifty-five years of age), those with ovaries undergo menopause wherein changes in the ovaries result in lowered production of estradiol and inhibin (a protein), which alters the production of the two main gonadotropins, follicle-stimulating hormone (FSH) and luteinizing hormone (LH), leading to the cessation of reproductive cycling.[50] Menopause has a variable range of physiological effects in addition to the changes in the hormone levels and is distinctive to humans (but possibly occurs in a few other animals as well). Individuals with testes undergo an "andropause" slightly later than menopause, with a decrease in the level of production of testosterone accompanied by a suite of physiological and physical changes.[51]

Patterned variation in hormone activity is an important aspect of sex biology. And, while there is much variation between individuals, there are species-wide typical patterns. Estradiol (a main form of estrogen) generally rises in individuals with ovaries during "mini-puberty" and then drops to levels similar to those in individuals with testes from ages two to eight. During actual puberty, estradiol rises in all humans but is usually two to three times higher in those with ovaries. As adults, those with ovaries, and who menstruate, typically have cyclical increases in estradiol during the follicular stages of the menstrual cycle. Progesterone spikes in all humans at birth and then drops to low levels until the age of seven or eight, when it rises moderately in those with ovaries and remains stable in those with testes. In adulthood, levels of progesterone are about the same in all humans, except during the luteal phase of those who menstruate, where it is up to twenty times higher, and during pregnancy, when it also increases. Testosterone levels are usually about three to ten times higher in infants with testes than in those with ovaries, but drop after mini-puberty, and all humans have about the same testosterone levels until true puberty. At puberty, all humans' testosterone rises, but those with testes typically undergo a larger spike. As adults, most humans with testes typically have circulating testosterone levels on average about ten to eighteen times higher than those with ovaries. However, there are a number of variants on this theme, both in humans with testes who produce very low amounts of testosterone and in variations such as PCOS (Polycystic ovary syndrome), which occurs in five to twenty percent of human adults with two X chromosomes and often results in high levels of androgens (especially testosterone) produced.[52] As with all hormones, the varying levels between bodies at different points in development and life history can have differing effects on various aspects of the developing body, physiology, and behavior. The patterns of hormone variation matter for human bodies, but they are not binary in nature or form.

One can't do a review of hormones without talking a bit more about testosterone. "T" holds an almost mythical role in the public view, and many think T is what "makes" a "man." But is that correct? T is probably the most scientifically researched, discussed, and debated, and certainly the most publicly misunderstood of the gonadal hormones. On the one hand, there are researchers who believe that testosterone's main job is to support the anatomy, physiology, and behavior that increases a male's reproductive output—T helps them [males] reproduce and direct energy to be used in ways that support competition for mates, in short, that testosterone is *the* biological and evolved "difference" between males and females.[53] Others argue that "T" is best understood in the context of a dynamic and malleable set of relationships. These scholars suggest that T is neither the creator of masculinity nor the essence of maleness, but is involved in growth, muscle-mass development, and reproductive dynamics that produce key patterns of variation in and between 3G bodies, and thus T is intricately related to the gender/sex experience.[54] These scholars also argue that T is often responding to, not necessarily driving, behavior and physiology. The preponderance of research demonstrates that testosterone action and effect is interlaced in and with biological, social, and cultural contexts. A full understanding of this (and other) hormone's role and impacts is not simply volumetric (how much is produced or is circulating), but has to do with the dynamics of receptors, feedback loops, patterns and variations of development, and a myriad of other physiological and social factors requiring an integrative, and nonbinary, approach.[55]

As for function, the entire current corpus of work on testosterone does offer some basic agreement: T is involved in aspects of muscle mass, aggression, and sexual behavior/desire in many animals, including humans—and not just in those with testes—but, then again, so is estrogen.[56] The distinctive evolutionary histories of different animal linages, like hyenas, naked mole rats, and titi monkeys, shape the specific patterns and dynamics of T and other hormones related to their sex biology in specific manners. Clearly, the distinctive aspects of human evolution altered some of the dynamics of T in humans. While the variance between testosterone levels, and some of its interactions/effects, can be, on average, substantial between those with testes and those with ovaries, within-3G sex correlation between testosterone and muscles, sex and aggression are small and often inconsistent. And, the patterns of T in relation to the wide array of people who do not fit into 3G categories are little studied or understood. Moreover, the interactions of T and parenting behavior in humans are pronounced, which is not surprising given the distinctive human evolutionary history around the caretaking of young.[57] At the end of the day, while testosterone is an important hormone with substantive sex-biology-related variation, it is not *the* marker of distinction between 3G categories or the "maker" of males or men.[58] The idea that testosterone—or any single hormone—is the biological basis of a sex binary is not supported by the research and reality of the complexities of the endocrine system.[59]

However, there is a suite of processes closely mediated by a diversity of hormones, sex biology, and reproduction, which occurs only in some human bodies and not in others.

Menstruation, Gestation, and Lactation: Now Here's a Difference

Unlike all we've covered so far in this chapter, there are three core processes that do mark clear, and distinct, biological differences between humans who have them and those who don't. These processes are central to mammalian reproduction and, as usual, the human lineage has done some weird things with them relative to other mammals.

Menstruating people and people who could menstruate make up half of humanity and thus a core aspect of sex biology.[60] Ovulation in humans typically happens on a cyclical basis in those with ovaries across a certain portion of the lifespan, between menarche and menopause. The menstrual cycle itself involves a range of physiological processes related to the preparation of the bodies of those with ovaries and uteruses for potential gamete fusion, possible pregnancy, and subsequent interactions with a zygote, embryo, and fetus. These processes are regulated by the hypothalamic-pituitary-ovarian (HPO) axis, specifically interactions between follicle-stimulating hormone and luteinizing hormone, progesterone, estradiol, and cortisol (plus a few others). Unlike much common (and unfortunately medical) belief, the menstrual cycle is not a closed system isolated from other aspects of one's body and life that is more or less the same across everyone who menstruates. Instead, it is a process that is impacted by and responds to many factors, including development, lived experience, age, nutritional state, activity, and so on.[61] The timing of first and last menstrual periods, the concentrations of hormones, and the thickness of the endometrium (the uterine lining) all vary with and are particularly responsive to stressors from the social and physical environment. As with so many key processes and traits in human sex biology, there is not one "normal" menstrual cycle or one "best" way for bodies to menstruate. And while only about half of humans can potentially menstruate, menstruation is a biological process, a cultural experience, and an important part of the entire human gender/sex landscape.[62] Even in an aspect of biology that only exists in a half of human bodies, an actual dimorphism, the dynamics of the process reject a simplistic description of the human experience of sex biology.

If one has ovaries plus the associated fallopian tubes and uterus, one can usually undergo pregnancy, which is the major physiological process involving the complex of gestation and birthing. Humans without a uterus are biologically unable perform this physiological process. Along with menstruation, pregnancy is probably the most substantive patterned difference between many in the 3G category of female and all 3G males. However, it is vital to note that not all humans categorized as 3G females can (or do) become pregnant, for a variety of reasons, and some humans who do not fit into the 3G categories can, and do, experience pregnancy.[63] There is also a range of physiological variation, including liver, heart, endocrine, and neurobiological changes, that can emerge between those individuals who have gestated and lactated and those who have not.[64] Given these patterns of variation, even in the experience of pregnancy humans do not map to a simple binary.

Pregnancy often results in an infant, and lactational feeding of infants is a significant aspect of human reproduction. While breast tissue itself emerges in most humans to some extent, lactation is possible due a specific process, postpuberty, typical in 3G-female individuals. In the process, estrogen, progesterone, and a few other hormones help facilitate the development of ducts, stroma, and glandular tissue with milk-producing globules that mix with the adipose tissue on the upper chest congregated around the nipples. In most humans with lactation-capable breasts who become pregnant, changes in prolactin, estrogen, and progesterone levels at the terminal end of pregnancy and after birth facilitate the maturity and activity of the milk-producing globules in the breast glandular tissue, enabling the production and expression of human milk. As noted above, breasts appear in as much as 30 to 60 percent of 3G males, but they are usually not capable of lactation.[65] There are rare cases of human 3G males exposed to high levels of prolactin expressing milk-like substances (galactorrhea—not true lactation). In some intersex individuals, lactation occurs, and it has been successfully induced in some individuals with XY twenty-third chromosomes and some transwomen who have undergone progesterone/progestin hormone therapies.[66] Intriguingly, when one looks across the mammalian world, there are two species of fruit bat where 3G males do regularly lactate[67] (via mammary glands with milk-producing globules) and feed offspring with their milk. While there are clear and typical patterns of variation for lactation (it's usually only in those humans who give birth), the biological variation across our and other species demonstrates that even in the key reproductive dynamic of lactation there is no simple binary.

There are two final sets of "internal" biology that deserve our attention, places that are the basis for deep and powerful assumptions about sex biology: the brain and DNA.

Brains/Neurobiology

Minds are a core to how we perceive, interpret, and respond to the world. And at the heart of the human mind is the human brain. There are some patterns of variation between the genders of men and women across human cultures in behavior, psychological illness, and the manners in which one perceives sights, sounds, and other signals from the world. This is often assumed to be due to biological "differences" in the brains of 3G males and 3G females. And, as usual, those humans who do not fit into the 3G categories are largely ignored in this area of research.

Are there male and female brains? A group of researchers reviewed hundreds of studies (conducted over more than thirty years) of human brain imaging and physical postmortem brain analyses. They discovered few patterned variants between 3G categories despite the strong history of repeated claims of "sex difference" in the human brain.[68] As with much in human bodies, 3G-male brains are on average larger than 3G-female brains from birth and stabilize at about 11 percent larger as adults. This size difference is often assumed to be distinct from the other commonly reported, but largely inaccurate, "differences" in brains: a higher white/gray matter ratio, more intra- versus interhemispheric connectivity, and higher regional cortical and subcortical volumes in 3G males. There is also some reported 3G-related variation in the connectome (a comprehensive map of neural connections in the brain), and there are relatively successful computer-based predictions of 3G sex of any given brain based on assessment of shapes and volumes of areas. However, all these patterns are primarily correlated with size differences, and they vary greatly across populations and individuals. Brain-imaging studies fail to identify consistent reproducible differences in how brains react to learning challenges (called activation differences) between 3G categories (and between genders) in verbal, spatial, or emotion processing. The largest recent overview of brain studies found only two small areas that are pretty consistent in their pattern of variation not due to overall brain size: subcortical structures called the amygdala and putamen. Also, the INAH-3 nucleus in the anterior hypothalamus appears to be relatively larger, on average, in 3G males. Overall, when the totality of findings on structural variation in human brains are assessed, the 3G-sex category explains only about 1 percent of total variation. This means that 99 percent of the structural variation in brains across humans is explained by factors *not connected* to 3G categories. Being "male" and "female" tells us very little about brain structure.

The conclusion of many researchers is that differences emerging from 3G-male/3G-female structural brain comparisons are largely trivial and often population-specific: the structure of the human brain is not sexually dimorphic.[69] This perspective and the overall data on brain structure and 3G sex are best summarized by a group of fourteen neuroscientists who state, "Although there are gender/sex differences in brain and behavior, humans and human brains are comprised of unique 'mosaics' of features, some more common in females compared with males, some more common in males compared with females, and some common in both females and males. Our results demonstrate that regardless of the cause of observed gender/sex differences in brain and behavior, human brains cannot be categorized into two distinct classes: male brain/female brain."[70]

There are researchers in the neurobiological arena who disagree that brains cannot be classified into female and male versions, arguing that there are 3G-sex variants in brain structure that are not due to average size differences. Shortly after the publication of the large overview noted above, a group of other neuroscientists published an analysis of age, 3G sex, and brain allometry (relative brain size)[71] using the UK Biobank dataset (of about forty thousand brains of primarily [95 percent] white British nationals) and a specific type of analysis software to retrofit the brain images for analysis.[72] This group found age differences and 3G-sex differences and age-by-sex differences in the volumes and shapes in older individuals (forty to sixty-five years of age).[73] While all their measures overlapped extensively across 3G categories, they did find some form of 3G-sex differences in 67 percent of the relative size measurements, with 37 percent of regions larger in 3G males and 30 percent in 3G females. They also found that 49 percent of all brain volumes generally decreased with age, and 28 percent increased, although aging effects varied across all categories of analyses. Importantly, the graphs of the data from the two 3G categories provided in the study show dramatically more overlap than the height graph at the start of this chapter, demonstrating that variation between all individuals in the dataset vastly outweighs the variation between 3G categories. Also, the main research group and subsequent commentaries on this topic highlight the fact that *there are, to date, no known corollaries of behavior or function* connected with reported relative volume/shape differences. It is not clear that any of these patterns do, or mean, anything at all.

In a commentary article published just after the UK Biobank study, another group of neuroscientists agree that there is key variation but argue that the patterns represented are not dimorphic, but are a continuum, with 3G categories exhibiting mean or variability differences (as with height, but much more overlap).[74] In other words, even for researchers who think these results demonstrate significant patterned variation, there are not two types of brain (male and female), but rather there is one range of brain function and 3G individuals are distributed along it with extensive overlap and some small differentiation. Interestingly, a study published a few years before on the same UK Biobank dataset, but a smaller sample and different methodology, also found some patterns of variation by 3G category, but they were different ones, and many were greatly reduced or disappeared with size corrections.[75] Finally, the researchers from the large overview presented above redid their key comparisons to include the UK Biobank study and five others studies looking at the same dataset or using similar methods and discovered that the results of all six varied in locations of variation, direction of the variation, and whether or not variation occurred in a given location in the brain.[76] The is no single, or same, pattern of variation across all studies. The current data and analyses demonstrate that there is little evidence for substantive structural differences across most of the brain, aside from those related to overall size.

This is not to say that brains in 3G-male and 3G-female bodies, those of people who do not fit into 3G categories, those of men and women, and those of very masculine or very feminine individuals, do not respond, or function, somewhat differently in response to specific stimuli, contexts, and actions; sometimes they do.[77] The fact that brain activity and the functioning of various aspects of brains varies by life experience, nutritional status, exposure to trauma, age, gender, degree of masculinity and femininity, and a myriad of other sociocultural and ecological factors, has been long known. Brains are biocultural.[78] This dynamic of little-to-no structural differences but a wide range of activity and a diversity of ways that the brain reacts suggests that rather than focusing on a search for "differences" between 3G categories, an approach that examines gender-influenced patterns and dynamics using a nonbinary frame is likely to be more beneficial and scientifically fruitful.[79] Even those arguing for clear and consistent structural differences acknowledge that "males and females are (at the group-level) often exposed to systematically different environments across the lifespan" and that this gendered reality could have systematic impacts on neurobiological activity and function.[80] The current data are pretty clear: there are not two brain types or even a continuous gradient from masculine to feminine; rather, the human brain appears to be a multidimensional "mosaic" of countless brain attributes that differ in distinctive patterns across individuals.[81]

DNA

I end this overview of contemporary human sex biology with the place most people want to start: DNA. Specifically, the first "G" in 3G: the genes on the twenty-third chromosome. This chromosome pair comes in two forms: X and Y. These chromosomes are often called the "sex chromosomes" because they contain many genes related to the development of gonads and other elements of the reproductive tract. But as the X and Y chromosomes themselves are neither sufficient to "produce sex" nor limited to directing the development of sex biology, the very heart of the supposed binary of sex is neither a binary nor just about sex biology.[82] The X chromosome has about 155 million base pairs of DNA and about 900 genes, not all of which are related to sex biology. The Y chromosome has about 59 million base pairs and about 55 genes.[83] Most humans are either XX or XY, but there are also many variations on this theme. Usually in individuals with more than one X chromosome there is a process of "X chromosome inactivation" where all but one of the Xs is "shut down" in each cell (not always the same X) so that its genes are not expressed. But this process is not always 100 percent effective, which can lead to mosaic development and a range of potential physiological effects.

Generally, if a human embryo has one X and one Y chromosome, genes on the Y chromosome "turn on" and facilitate the development of testes and related physiology. If one has two Xs, then the active genes facilitate the development of ovaries, uterus, and related physiological patterns. These developmental trajectories primarily target reproductive physiology, but also have downstream effects on patterns of hormonal action, muscle, and bone development.

But being XX or XY does not always correlate with the patterns just described.[84] One can be XY and have a variation in the activation of segments of DNA so that the specific genes that facilitate testes development never turn on and/or their protein products are differently made or transported. The same is true for someone who is XX and their ovary-related genetic sequences. Variations that depart from a 3G pattern of gonad and genital development are rare, but as noted earlier, they are a consistent aspect of human biology. For example, the SRY gene on the Y chromosome is assumed to be a TDF (testes determining factor) and known to steer the developing embryo toward the formation of testes. But it can be present and not "turn on" or sometimes can have been translocated to another part of the genome. There are also a number of other genes shared by all humans, being on the X chromosome or one of the twenty-two autosomes, such as DAX1, SOX9, and Wnt4, that all appear to have dosage effects (when there is too much or too little of a gene product) that affect the patterns of gonad and genital development.[85] Variation in the activity of these genes in embryos affects much of the variation in gonads and genitals in 3G-sex individuals and in those who do not map perfectly into the 3G categories.

There is a wide array of other variations on the dynamics of the twenty-third chromosome, such as the occurrence of XO (no Y or second X), XXY, and XYY individuals. There is also a variety of scenarios where physiological processes are impacted by more than just gene action in the development of an individual and facilitate a range of outcomes that challenge the 3G-sex category model, such as XX individuals with a penis, XY individuals with a clitoris, vagina, and labia, and individuals who are XX or XY but have mixed sex gonads, genitals, and other variants. Overtly non-3G bodies make up a very small proportion of all humans, but many more covert variations arising from these same dynamics give rise to a range of genital, gonad, and endocrine structures and patterns that are outside the typical 3G expectations but are still loosely classifiable into 3G categories.[86] Such variation is understudied and poorly understood but is a recurrent aspect of human biological processes and reflects the normal range of variation in reproductive development.

Finally, recent work on the human transcriptome (the initial biochemical products of genes) demonstrates some measurable variation in initial products by identical genes in different tissues of the bodies of 3G males and 3G females. However, the end product or action that these genes are associated with is almost always the same across all bodies. This is important because even when genes are doing some things slightly differently in 3G-category bodies, they mostly end up with the same outcome.[87] There is, however, a specific subset of the genes that act differently depending on which specific tissues associated with 3G categories they are expressed in. For example, there are about 6,500 genes that show some 3G-sex-based variants in their transcriptome action. About 6,000 of those variants are related to expression in mammary tissue, and the majority of the rest are in reproductive related tissues, with a few in adipose, skeletal, and muscle tissues—all of the specific areas where endocrine, growth, and structural variations affecting sex biology are the most prominent—and all measured in adults.[88] This intriguing genetic data suggest that there are some dynamic relationships between growth/development, endocrine and reproductive physiology, gene expression, and the lived experiences of humans based on aspects of sex biology. Yet again, even at the level of gene functions, there are dynamic and variable effects and processes, but there is no sex binary.[89]

Normal?

The actual biological variation of any human trait comes with a set of ideas about what constitutes a "normal" version of that trait. This idea of "normal" is also almost always accompanied by opinions about what that trait *should* be like. But this should give us pause and makes us ask the question: what is "normal" human biology?[90]

The average measure in a range of distribution of a given trait is called the "mean" of that range. It is not necessarily the best or ideal form, nor is it necessarily the most functional, adaptive, or even aesthetic. It is a statistical representation of the sum of the range of values in a given measure divided by the total number of values in the dataset. In many of the traits related to sex biology that vary across human bodies, the mean has a cultural value, and it is thought to be "normal." And people are judged against it. Therefore, when asking about aspects of sex biology, the range of variation and functions or lack of function associated with that range should be the focus, not necessarily the mean or some subset of the typical distribution. Comparing means of some traits, like height and hair, have far more cultural meaning than biological significance. Other traits, such as gonad type or specific hormone levels, can have equally salient cultural and biological significance, but the two might not be connected in a simple fashion. Still other traits, like muscle strength, vary in their salience and importance depending on what specific questions are being asked and what comparisons are being made. In the reality of human sex biology, means, ranges, and traits have both cultural and biological implications that are almost always entangled.

Ultimately, there are some important patterns of variation between 3G categories, but there is a much more overlap in functions and outcomes than can be captured by the concept of sexual dimorphism or "sex differences."[91] "Female" and "male" do mean something biologically but are not two different kinds.[92] X and Y chromosomes, gonads, pregnancy, circulating hormone levels, genetic processes, developmental dynamics, musculature, hair growth, adipose deposition, and brain morphology and function all impact the human experience. But these aspects of our biology, and their variation, are not best understood as binary sex differences. Humans are biocultural and have gender/sex. Variation is our norm, and our variation does not reflect two kinds of human.

6

No Biological Battle of the Sexes

IN 1993, a book called *Men Are from Mars, Women Are from Venus* hit the bestseller list and stayed there for 121 weeks. In it, the author and relationship counselor John Gray claimed that men, like the Roman god of war, Mars, are by nature aggressive and violent, but protectors, and women are emotive, coy, and maternal like the goddess of love, Venus. He told the readers that successful male-female romantic relationships are based on recognizing natural (biological) differences in communication, emotion, and behavior between men and women.[1]

Sound familiar?

Charles Darwin told us that "man is more courageous, pugnacious, and energetic than woman" with "more inventive genius."[2] Angus Bateman claimed males are promiscuous and females sexually passive, coy, and choosy, and E. O. Wilson told us that the evolved, biological differences between men and women naturally led to universal male dominance.[3] Too many biologists and other kinds of scholars across the nineteenth and twentieth centuries made these kinds of assertions. And these assertions, just like those of John Gray, are wrong.[4] We know that 3G females are not necessarily sexually passive and that 3G males are not necessarily more courageous or energetic, nor have humans evolved for male dominance or two different kinds of sex personality. Yet, more than three decades later, *Men Are from Mars, Women Are from Venus* is still selling tens of thousands of copies, and some biologists still refer to sex as a binary, imagining that a male/female split, via anisogamy, pits the sexes against one another and explains much in the human experience. The commitment to the binary view and its "battle of the sexes" implications is deep and resilient. Given all we know, this should not stand. So, yet again, let's take a closer look.

It's true that many species have sex-biology patterns that lead to marked distinctions in bodies and lives between sex categories. Take, for example, peafowl (peacocks and peahens). Peacocks, the small-gamete producers, have much larger bodies, extravagant coloration, and enormous tail feathers, potentially creating different modes of movement, risk, and related behavior relative to the peahen, the large-gamete producers. But humans are nothing like peafowl. We are not a species with dramatic, evolved patterns of sex difference.

The core premise of the battle of the sexes (in mammals) is that in order to successfully reproduce, the 3G female is locked into gestating and caring for infants and the 3G male is wired to get as many copies of his DNA (via sperm) into as many 3G females' reproductive tracts as possible. The classic anisogamy position in biology asserts that there is a vast disparity in energetic investment in gamete production and investment in offspring between 3G males and 3G females that is related to differences in reproductive organs and physiology. As a result, there is an evolution of distinctive anatomical, physiological, neurobiological, and behavioral processes due to differential evolutionary pressures on the sexes.

When it comes to humans, the biology used to "demonstrate" the validity of a battle of the sexes is body size and strength (larger and stronger 3G males) and specific behavioral and hormonal differences (especially more testosterone in 3G males). For example, some researchers assert that "sex differences in muscle and strength are largely the result of sexual selection for male-male competitive traits, whereas females have increased body fat (in place of muscle) due to natural selection for maternal investment capacity in the context of our unusually large brains."[5] These researchers believe that evolution has shaped human bodies along two distinct lines of development, producing 3G males geared to fight each other for females and 3G females shaped for raising young. Chapters 4 and 5 outlined why these assumptions don't map onto what we know about human evolution and how they oversimplify descriptions of the existing variation in human biology. Along these same lines, others argue that it is specific patterns of circulating levels of testosterone that create massive 3G-male and 3G-female bodily differences and that "T" even "pushes the psychology and behavior of the sexes apart."[6] In the last chapter, we covered how this is an oversimplified assignment of superpowers to one of the dynamic, and variable, aspects of the human endocrine system.[7]

So, as we see, there are scholars who argue that human 3G-category bodies and minds are deeply and evolutionarily different and that this "biological reality" places human men and women in direct conflict. And there is a large swath of the public that believes this as well. However, scientific data and analyses demonstrate that this sex-conflict position is not true.

Of Minds, Means, and Behavior

The "females and males as very different" camp often points to cognition ("minds") as the proof of their position. Everyone knows men and women think differently, right? Over the past few decades, massive studies called meta-analyses[8] reviewed patterns in many of these cognitive variables, such as math, verbal, and spatial-ability skills, communication dynamics (verbal and nonverbal), social and personality variables such as aggression, negotiation, helping, sexuality, leadership, introversion/extroversion, general psychological well-being, some motor behaviors (throwing, balance, flexibility, etc.), and a few other psychological states and behaviors (moral reasoning, cheating behavior, etc.). These meta-analyses involved data from more than twenty thousand separate studies involving more than twelve million participants.[9] The results are clear: Across most topic areas in psychological science, the difference in responses and outcomes between males and females is small or very small.[10]

Here "small" and "very small" are measures of how far apart the means of massively overlapping variation are. That is, pretty much everything being measured in these studies overlaps almost completely between 3G categories (usually based on self-reported genders), but the means of the distributions of the measured variables, when separated by 3G-sex category or gender, can be different from one another. Think of the height example from earlier in the book: 78 percent of folks in the United States are not identifiable to 3G sex simply by height, but the means of the overlapping distributions of 3G-male and 3G-female height are different. So, on average, one can say 3G males are taller than 3G females. But that might not tell you much at all about any specific individual, or about height as a biological characteristic, given the massive overlap between 3G categories. In the "mind" studies, most of the differences between the means are much, much smaller than in the height example. The difference between the means in these studies is assessed by the common statistical tool called the "cohens *d*" measure, which reflects how far apart the means in the overlapping distributions of the measurements are in standardized units. So "small" suggests that the means are very close to one another and "very small" even closer (with almost 100 percent overlap). The between-gender differences in these huge meta-analyses were small in 46.1 percent of all cases and very small in 39.4 percent of all cases. The largest and most recent metanalysis demonstrated this same massive overlap, with about 84 percent of mean differences being small and very small across most traits examined.[11] Again, we really aren't so different. And cognition is certainly not binary.

While the vast majority of the traits measured in these studies overlap extensively, there are a few patterns worth noting. Men (the self-identified gender) consistently scored higher in hand-grip strength, sprinting, throwing velocity and throwing distance, masturbation, views on casual sex, physical aggression, and mental rotation of objects. Women (the self-identified gender) scored higher on indirect aggression, agreeableness, and smiling. These patterns reflect the effects of variation in muscle mass and body size, training, and patterns of enculturation (gender). But none of them stand out as clear and unequivocal markers of evolved patterns of male-female conflict.

The patterns of higher scores for casual sex and masturbation in men often receive a lot of attention, and are used to argue for an evolved 3G-male reproductive strategy of "more sex, less investment."[12] However, self-reporting on sexuality and sexual experience is notoriously problematic (e.g., men overreport and women underreport) due to the powerful gendered expectations and cultural constraints in the contemporary world.[13] Even if the reports were mostly accurate and self-identified men do indeed have a greater interest in casual sex, more sex partners (on average), and more frequent masturbation, it would not signify that "man's mind" differs from "woman's mind" due to evolved patterns. One would need to have a comparative sample without the contemporary global cultural views on sex, sexuality, and patriarchal social structures to control for such variables. Also, from a purely biological/anatomical perspective, the wholly external location and the physical structure of the penis and testes makes it easy to manually manipulate to stimulate the sexual physiological response (e.g., masturbation). While the clitoris is also external and highly responsive to stimulation, across the human societies where the studies on sexuality are done, there are substantively different beliefs, expectations, and restrictions on sexuality, including masturbation, for those with clitorises as opposed to those with penises. This likely plays into the self-reported differences in sexual behavior and self-sex.

In mammals with hands, especially primates, one often sees individuals with penises engage manual stimulation for its physiological reward (primates masturbate a lot). The increased frequency of reported masturbation in humans with external penises and testes is at least as likely a by-product of the combination of primate hands and genital anatomy in the context of specific human cultural systems, as it is a reflection of an evolved difference in mind and body between those with and without the penis/external testes combo. Similarly, the lower self-reported rates for human females may not be due to any biological difference, but rather to cultural ones. In many primate species (including humans), females *do* masturbate, and do so more than in most other mammals. Female primates are also observed to use objects to masturbate. In fact, research on monkeys and apes demonstrates that "females exhibit more sophisticated kinds of masturbatory behaviours than males."[14] Primates use sexual behavior outside of reproduction a lot, and masturbation is part of that. Humans are typical primates in this respect (see below); it's just that for us sexual behavior also has substantive cultural connotations and consequences. And these cultural realties in much of the contemporary world often restrict the expression of sexuality by those classified as female. Thus, measures of sexual behavior need to take that into account. As the biologist Anne Fausto-Sterling has eloquently and repeatedly demonstrated, "Sexuality is a somatic fact created by cultural effect."[15] In other words, sexuality is not strictly programmed by our genes, or genitals, but rather emerges in concert with the gender/sex experience of humans as they develop.

Actual data on sexual activity are difficult to collect and quantify, but one can look to the 2010 National Survey of Sexual Health and Behavior, a US-based study of 5,865 adolescents and adults (2,936 self-identified men [gender] and 2,929 self-identified women [gender] ages fourteen to ninety-four) to get a general idea.[16] In the study, 55 percent of the men reported masturbation in the past month, and 71 percent in the last year; 31 percent of the women reported masturbation in the last month, and 54 percent in the last year, except those over seventy. 85 percent of men in their twenties and thirties reported having vaginal intercourse in the last year, compared to 74 percent in their forties, 58 percent in their fifties, 54 percent in their sixties, and 43 percent in their seventies. For the women, 81 percent in their twenties and thirties reported having vaginal intercourse in the last year, compared to 70 percent in their forties, 51 percent in their fifties, 42 percent in their sixties, and 22 percent in their seventies. Men and women of all age groups reported engaging in oral sex and masturbation with a partner. For both oral sex and partnered masturbation, the man/woman pattern is almost identical. More than 20 percent of the men between ages twenty-five to twenty-nine reported anal sex, with younger and older men reporting much lower numbers. More than 40 percent of the men eighteen to fifty-nine years old reported participating in anal sex during their lifetimes. Women are almost identical with slightly higher frequencies of anal sex over a larger age range (eighteen to sixty-nine) than males. Across all age categories, 8–10 percent of the men and 5–9 percent of women reported same-sex sexual activity during their lifetime with much higher figures (up to 17 percent) for the twenty to thirty-nine age group. The survey did not ask about sexual orientation, so it's unclear what percentage of these numbers reflect homosexually oriented individuals as opposed to heterosexual or bisexual persons engaging in same-sex sexual behavior.[17]

While this is just one example, a plethora of studies of human sexual behavior demonstrate that much human sexual activity is outside of reproductive possibilities; humans have sex across a wide range of contexts and in multiple manners and that there is a lot of variation across individuals in sexual activity, orientations, desires, and practices.[18] One pattern that emerges is the overall decline in sexual activity with age (especially over sixty). However, women report a higher decline than men. This outcome may be related to physiological changes associated with the sex biology of menopause.[19] It also might be related to patterns of cultural heteronormativity and patriarchy that encourage older men/younger women pairings. Given everything known about human sexuality, it is most certainly biocultural.

Overall, the data show few if any consistent patterned differences between human men and women in actual sexual activity. Regardless, many still argue that the real differences between sexes are not in actual sexual activity but the patterns of interest in sex as shaped by evolved patterns of bodies and lives based on assumptions about anisogamy (whether you produce small gametes or large gametes, etc.). This pattern of differential interest in sex is assessed via a person's "sociosexual orientation," which is measured via the Sociosexual Orientation Inventory (SOI). The basic argument is that there is a pervasive pattern of evolved differences between males and females, and that this can be seen in their attitudes about sex in general, sexual fantasy, and sexual behavior.[20]

In general, men (the gender) tend to score higher than women (the gender) on the SOI. Men also tend to report higher interest in sexual activity and in sexual fantasies, higher numbers of preferred or actual sexual partners, and prefer short-term versus long-term sexual opportunities (on average). There is also a robust degree of self-reported variation cross-culturally (across genders and 3G categories) in SOI-related patterns.[21] However, there are no neurobiological patterns or aspects of reproductive physiology that indicate such patterns come from specific bodily variation in 3G males and 3G females. Even testosterone, which plays an important role in sexual arousal in our species, does so in all humans not more or less in one 3G sex. There is a set of well-studied physiological links between testosterone-level fluctuation, sexual activity, sexual bonding, and parenting in some groups of 3G males.[22] Unfortunately, there is not nearly as much study of the specific dynamics of T in 3G-female sexual activity and sociosexuality as there is in 3G males, and almost no study of this at all in people who do not neatly fit into 3G categories, so truly scientific comparison is difficult.[23] Given the substantive average variation in circulating T levels between 3G females and 3G males, and the on-average patterns of SOI difference, one might think that the assertion that higher T in 3G males is related to (but not uniquely causal of) the between-category variation reported in sociosexual orientation is accurate. However, it is possible that circulating T isn't the right measure to assess with regards to sociosexuality. Above a low threshold, there's little to no increase in libido with an increase in circulating T among most 3G males, and smaller increments of T seem to have libido effects in 3G females. So, rather than simply circulating T as the prime agent that underlies libido, it might be T-receptor variation, and a range of elements related to gender and cultural dynamics also in play.[24] The main researchers who study sociosexuality argue that T is not the only or even the primary factor, stating that "the most consistent finding was that men scored higher than women on sociosexuality across cultures. Several different theories were evaluated concerning why men and women differ in this way. They all received at least some empirical support. As a result, we are left with the relatively unsatisfying conclusion that sociosexual sex differences are predictable from several theoretical perspectives, none of which is conspicuously superior to the others. . . . At present, it appears that multiple perspectives are required to more fully explain the cultural and gender-linked variance in sociosexuality."[25]

There is some gender/sex-related variation in self-reported perspectives on sexuality, but this variation is not simply a reflection of an "evolved difference" stemming from gamete sizes and reproductive tracts. Humans have a lot of sex (more than most other animals), and there are relatively few, if any, significant biological features or variants that constantly differ between 3G males and 3G females in relation to kinds and patterns of actual sexual activity. Humans are biocultural creatures, and the social and cultural context in which we develop is going to have enormous influence on something as complex as our self-reported perceptions of sex and sexuality.[26]

Aggression is another area where patterns of variation emerge. It is also an area that has been held up as an indicator of evolved male-female conflict.[27] On average, 3G males have a larger body size and greater upper-body muscle mass and therefore can pose a potentially greater risk of harm in purely physical fights than do smaller 3G females (assuming no training in hand-to-hand combat). Across many societies, young adult men (the gender), on average, tend to participate in some culturally defined categories of physical risk-taking behavior at higher rates than young adult women (the gender).[28] In many societies, men engage in overall higher rates of physical aggression than women, especially in conflicts with other men. But in aggression between heterosexual partners there is very little difference in the use of physical aggression, with women being slightly more likely to use it.[29] However, in this same context (heterosexual partnerships/pairs), physical aggression by men often causes more injury than that by women. This is not surprising given that 3G males are on average slightly larger than the woman they are fighting with, and often have a greater upper-body strength due to both physiological and cultural patterns, and are likely to have more previous experience in physical fighting. This outcome is a biocultural blend of patterns, not a smoking gun for evolved differences in physical violence between 3G categories.

There is a particular consistency in certain reported patterns in aggression and violence, especially at the level of homicide and sexual assault/rape, when using the 3G-male/3G-female comparison.[30] Some have argued that this is related to the differences in circulating testosterone and/or the evolution of violence and aggression as a basal adaptation of 3G-male behavior. However, there is no solid evidence from the human evolutionary record or human physiology to support such an assertion.[31] While 3G males are indeed the initiators of the vast majority of homicide and sexual assault/rape in our species, it is also true that most 3G males are never involved in such acts.[32]

Homicide and sexual assault/rape are, in most cases, not driven by reproductive outcomes but are acts of social violence, power, and control. Acts not reflective of specific evolved sex-biology patterns or traits but rather events catalyzed, or facilitated, by cultural histories, dynamics, and structures in the context of human gender/sex.[33]

Do patterns of sexual violence and homicide tell us something about 3G males evolving to be more aggressive than 3G females? Not really. Men (the gender) appear to be more aggressive in some contexts and women (the gender) in others. Men's physical aggression and men's participation in antisocial aggression (e.g., violent crime) are more common and often more serious than those of women. But the majority of all humans (men and women) do not engage in violent aggression, or crime, with any regularity, or even at all. Overviews of patterns in contemporary human societies, and of human societies of the past, and in-depth studies of human bodies in the past and today (chapters 4 and 5) strongly suggest that while there *are* gendered patterns of behavior that matter, there is no evidence for an evolved pattern of specific 3G-male hyperaggression as a key adaption in humans.[34]

As for brains and minds, many researchers conclude[35] that male/female physical/structural brain differences are largely trivial and often population-specific: the structure of the human brain is not sexually dimorphic, and while some features are more common in 3G females than in 3G males and vice versa, there are no "male brains" and "female brains."[36] There is not even a continuous gradient from men to women or masculine to feminine; rather, the human brain is a multidimensional mosaic of myriad attributes that differ in unique patterns across individuals.[37] There is patterned variation in the measured means for some mental states and cognitive behaviors in humans, but these are almost always small or very small and usually in the midst of massively overlapping variation. A focus on means obscures the likely more biologically relevant realities and dynamics of variation in human bodies and minds. If one seeks to validate assumptions about difference by looking only for difference, then that is all one sees. The patterns of variation in a forest are often more important than the average height of certain types of trees. The same goes for humans and human societies.

When the great overlap (and complexity) in sexuality, aggression, and cognition fails to support the battle of the sexes argument, its proponents often turn to the costs of reproduction as a trump card. But as we already know, scientific evidence about human reproduction and the evolution of human caretaking systems rejects the simple "large-gamete producers versus small-gamete producers" view of the world. To summarize: parental investment in humans is more complex than one 3G male and one 3G female investing in reproduction. There is massive evidence that the genus *Homo* evolved complex cooperative caretaking of young early in the Pleistocene and that by the mid-Pleistocene human childhoods had extended substantially because of it.[38] The evolutionary implication of this reality rejects a simple calculus based on individual 3G females' costs of gestation and lactation and includes a caretaking regime where the mother is *one of many* caretakers and the range of the specific infant caretaking energetic burdens is redistributed across the group. This is not to say that the "costs" of reproduction are not high in the genus *Homo*, but rather that the costs have, in an evolutionary response, been socially and physiologically mediated and spread across group members to enable the complex, extended childhood associated with a range of capacities connected to substantial neurobiological and social development after birth.

The nuclear family structure is often brought up as evidence for some aspect of the conflict between or severely different roles for 3G males and 3G females (male the provider and female the caretaker). The assumption is that the core nature of the nuclear family deep in human evolutionary history structures both *Homo* bodies and societies. However, as we reviewed in chapter 4, there is little fossil or comparative evidence of two-adults-plus-offspring as the core social structure for *Homo*. That is, existing fossil and archeological data do not offer clarity regarding group structure aside from the high likelihood of *Homo* groups consisting of multiple adult females and males and young. Whether the within-group structure consisted of multiple bonded pairs is not detectable, but having pair bonding as a common facet of *Homo* social organization is likely given the near-universal human social behavior pattern of pair bonding and its physiological manifestations. Pair bonding does not equal monogamy or two-adult exclusive mating or a specific two-adult caretaking pattern; it is not equal to or indicative of a nuclear family structure.[39] In other words, the current fossil, comparative, and contemporary data offer no support for the nuclear family as the "traditional" or core evolutionary basis of the human family or social group.[40]

Man ≠ Male and Woman ≠ Female: Beyond the Battle of the Sexes

The "battle of the sexes" assumes that man is interchangeable with male, as is woman with female. But this is not necessarily the case. Yes, 3G males and 3G females are typical clusters of genes, gonads, and genitals, and they vary in some patterned ways, but one's 3G state, typical or otherwise, does not uniformly tell us what one's gender identity, one's degree of masculinity or femininity, one's sexuality, or one's sense of self in regard to reproductive roles is or will be. And these 3G categories often tell us even less about one's other behaviors. Also, there is a cluster of humanity who are not easily classifiable into 3G categories and who need to be included in all these analyses if we wish to cover the full biological range of being human. As a collection of researchers in the areas of sex and gender recently noted, "Gender/sex is multidimensional and each component is dynamic and responsive to both internal forces (biological, cognitive) and external forces (social interactions, culture). Individuals show variability across the different components of gender/sex, presenting a mosaic of biological and psychological characteristics that may not all align in a single category of the gender binary."[41]

Given everything we've covered so far, it's clear that the human "sexes" are not just biology or just culture, they are not "the same," but nor are they "different kinds." Rather than envisioning typical reproductive physiologies, bodies, and minds in competition with one another (the "battle" of the sexes), the goal should be to move beyond assumptions about conflict between and uniformity of "sex" as categories and shift toward examining the actual patterns and variations in *Homo sapiens* and allowing the information emerging from that process to structure the inquiry. Male/female, man/woman, small gamete/large gamete, ovaries/testes, penis/vagina, and so on are not necessarily default comparisons to use when thinking about people. Of course, such comparisons can be appropriate for many questions, but we need to be open to letting actual behavioral, physiological, genetic, cultural, and historical data and variation drive the modes of categorization for our investigations, not cultural assumptions about what sex "really" means.

To be clear, there are patterned variations in sex biology in human bodies, and they contribute to the shaping of peoples' lives. But those patterns are not ubiquitous. They are seldom truly dimorphic, almost never binary, and do not always or usefully match to the categories of man or woman. In fact, close attention to the actual variation in caretaking, sexuality, gender identity, parental investment, aggression, and other behaviors might suggest other ways to cluster humans as the targets of the research focus and thus frame the question(s) differently.

Man is *not* more courageous, pugnacious, energetic, or inventive than woman. Men are *not* biologically more promiscuous than women, who are in turn passive and coy. Evolved, biological differences between men and women do *not* naturally lead to universal male dominance. And certainly, men are not from Mars and women are not from Venus. We are all the same species from the same planet. However, this does not mean that there are not conflicts between gendered expectations and experiences, that there is no oppression and exclusion between and within genders, especially in the context of patriarchal systems, or that reproduction, child care, and patterns of violence are equitably spread across all members of our species. What is important to understand is that the vast majority of the patterns and processes related to gender/sex that humans experience today are not biologically predetermined, evolutionary ordained, or inevitable. They are biocultural, created by the entanglements of human bodies, histories, lives and cultures. And thus they are not necessarily fixed or permanent aspects of the human experience.

7

Why the Binary View Is a Problem

TODAY, THE STATE of biological knowledge is one that "acknowledges sex is a rich, various, and diverse phenomenon that can—and should—be measured across multiple levels of biological organization and can be variable within an individual, within a species, and across different species. The study of sex diversity and variability in the animal kingdom has been hindered by imposing binary assumptions and limitations on what sex is, or can be, across species. By simply acknowledging that sex can, and does, exist outside a strict binary framework, we can evolve and improve how we define, measure, and analyze 'sex' in our research."[1]

Despite this knowledge, a binary perspective remains pervasive. Understanding and communicating the realities of human sex biology is not just an academic or philosophical undertaking; allegiance to the binary view in industry, government, education, and by the public fosters ignorance, harm, and suffering. Humans are always immersed in complex, variable, and biocultural gender/sex realities. Variation in human bodies, cultures, and lives illustrates that "average," "ideal," "typical," and "binary" are seldom the most effective ways to talk or think about human gender/sex.[2] This knowledge is powerful, but only if it is shared and used. In that spirit, this final chapter offers brief examples of a few areas in contemporary society where binary views of sex biology and gender are invoked, and it illustrates why such actions are both wrong and harmful.

Sexuality and Sexual Orientation

Back in 1991, a neuroscientist named Simon LeVay argued that a certain brain region, called the INAH-3, was less than half the size in gay men than in straight men, and that gay men and straight" women had INAH-3 of about the same size.[3] He believed he'd found the biological marker of being gay, in men. LeVay worried what society would do with the knowledge that "gayness" is located biologically. In discussing the possibility of a "gay" gene, he said, "We are going to have this power over our own natures, particularly over our children's natures. What we do with these choices will be one of the major ethical questions of this century. Are parents going to have free reign, or should society step in?"[4] Luckily, there will be no manipulation of embryos for sexual orientation, as subsequent research demonstrated that there is no "gay" area of the brain, nor are there "gay" genes. In fact, recent work reveals an incredibly complex and dynamic suite of genetic associations with sexuality and sexual orientation and demonstrates that the neurobiological, behavioral, and social relations to sexuality are equally complex.[5] The quest for a "gay brain" or a "gay gene" stems from the belief of a "natural" sex binary tied to heterosexual reproduction such that any significant deviation from that pattern, like homosexuality, must come from a biological alteration or aberration. But reality is far more complicated. Of course, there are biological components to sexuality and sexual orientation, but these are not simply tied to a specific aspect of sex biology.

There is an enormous literature examining and debating which biological and social categories or variants matter most for sexuality and sexual orientation, with little agreement aside from the fact that it is complicated.[6] The state-of-the-art understanding is that there is no consistent, or effective, sole predictor or single determinant for sexuality and sexual orientation in humans: not one's gender, nor one's genitals, nor one's gonads or hormones, nor one's genes or brain structures. Many people assume that heterosexuality is the "natural" mode of sexual behavior in humans and that other modes of sexuality, homosexuality, bisexuality, asexuality, and so on are either aberrations or deviations from the typical pattern. Such an assumption is a misleading way to discuss human sexuality. The physiological system of arousal is almost identical across all humans (regardless of what kinds of genitals are in play[7]), but the psychological, social, and emotional triggers for such arousal vary widely, are related to gendered identities, and are not solely, or even primarily, based on the specific variants of sex biology one has.

There is variation between, and within, gender and sex categories in sexuality. The sociosexuality index (SOI), initiation of sexual violence, self-reporting of masturbation, and a range of other behavioral patterns vary between and within genders and 3G sexes. For example, we noted in the last chapter that decline in interest in sexual activity seems to be affected by variations in sex biology related to age, relationship longevity, gender, and sexuality. There is a general decline in interest in sexual activity with age (especially over sixty years) in humans, but the decline for self-identified women is larger. This trend is especially acute in married or long-term couples (both hetero- and homosexual), where women's participation in sexual activity with their partners is negatively correlated with the length of time together.[8] This difference might be influenced by physiological changes affecting genitals and physiologies in bodies with uteri and ovaries during menopause combined with other social and psychological dynamics of aging. It also may be influenced by cultural structures and patterns, such as patterned partner-age differences, expected sexual roles and behavior, societal views of beauty, bodies, and sexuality, and a range of structured gender inequities in contemporary societies.[9] As with most things, there is not a uniform pattern; rather, there is only an "on average" one, and it is surely biocultural—disentangling the role of sex-biology variation from the cultural and gender context might not be possible in this case (or even in most cases of sexuality).

Sexual behavior varies substantially across many animal species, especially social mammals. In primates, sexual behavior of all kinds is more common in social contexts than in purely reproductive ones. That is not to say that reproductive sexual activity, which is by definition heterosexual in mammals and birds but not necessarily so in other animals (remember worm sex biology), is not common—indeed, it is typical. But the key point here is that the range of sexual activity, and pair-bonding, in humans is not necessarily constrained by (or even related to) the necessities of reproduction. Humans exhibit the most diversity and highest frequency of sexual behavior of the primates (except for perhaps bonobos), and all human cultures demonstrate a wide range of sexual behaviors and sexualities, with varying patterns of cultural restrictions and constraints. So, for humans the accurate characterization of sexuality and sexual behavior is that it is diverse, largely social, and often occurs across multiple partner combinations across human lifetimes in the context of a range of cultural rules, expectations, and institutions. One cannot scientifically look exclusively to sex biology or any biological specifics to fully understand any given human's sexuality and sexual orientation. Representing human sexuality as a binary system (men are one way and women are another, male/female, gay/straight, etc.) sets up erroneous and scientifically unsupported understandings and expectations regarding sexuality and sexual orientation. The binary approach tells us very little about human experience, and it likely obstructs effective and important research into sexuality.

Making a Family

In 2020, the Republican governor of the US state of Tennessee, Bill Lee, signed a bill to ensure that foster care and adoption agencies could exclude LGBTQ families and others based on "religious beliefs."[10] By 2024, there were at least thirteen states in the United States that had similar laws.[11] The laws are based on false beliefs about what is biologically "natural," and therefore "right" for humans. These laws assume that only heterosexual pairings of certain types of people (cisgender) provide appropriate and successful conditions for the rearing of children. This is factually incorrect, as the children of LGBTQ parents fare just as well as children of non-LGBTQ parents.[12] But these beliefs and real legal structures are not about the data or facts; rather, they reflect a deep commitment to the logic of the sex-biology binary: if female bodies are made for reproduction and caretaking, and male bodies for protection and provisioning, then the biological (natural) basis for family is a heterosexual male and female couple.

The core of mammalian reproduction is gametic fusion in a body capable of gestating and birthing. But the necessary biological details for the creation and gestating of an embryo tell us only a small slice of the dynamics of a human family. The human family is not limited to or by the physiology of reproduction. Humans create and nurture families, communities of care, wherein adults, and others, collaborate to care for each other and young. But families come in all shapes and sizes. The idea of the heterosexual nuclear family as basal to humanity is refuted by a range of scholarship.[13]

Humans with uteri play a central, biological role in the creation of offspring (gestation and lactation), and thus are often at the heart of the human family experience. Birth mothers can technically undertake much of the reproductive process alone, but historical, social, and biological features of humanity indicate that the human system of reproduction and caretaking has evolved to be a multi-individual process. Humans with uteri are center stage regarding gestation and birthing, but human-wide physiology is structured to enable caretaking of infants and strong bonding by individuals across age groups regardless of which variants of sex biology one has. Successful rearing of children and the formation of communities of bonding and care (families) occur in a variety of ways across a variety of humans. Unfortunately, there are some contemporary societies with structures, and laws, as noted above, that seek to inhibit or limit the diversity and possibilities in human patterns of family formation, community care, and childrearing by mandating that only one combination of humans (one 3G male and one 3G female) is the correct nucleus of a family. Given what we know about sex-biology variation, pair-bonding, family formation, and how the caretaking system in humans evolved, these laws, and beliefs, place a wide range of people, especially the children who benefit from the diversity and commitment of caretakers, at substantive risk. Human families are not necessarily composed of only biological kin, are not necessarily based on the union of two cisgender different 3G-sex adults, and they are not only about reproduction. Families are complex relations and networks of bonding, caretaking, and collaborating between socially close individuals. They are not the constrained product of, or necessarily related to, a sex binary.

Medicine

Historically in the medical sciences, researchers were often hesitant to use female animals in laboratory studies. Male lab animals (mostly mice) outnumber female ones by a factor of at least five, especially in tests focused on obtaining species-wide results.[14] Traditionally, medical researchers considered female animals, including humans, too behaviorally and hormonally unstable and physiologically complicated to give generalizable results in many cases.[15] In the medical binary view of sex biology, there are two types in every species, the male and the female. The male is the standard, the stable specimen, because his only contribution to reproduction is sperm. The female, with her instability, is shaped by her possession of a uterus and her reproductive cycling; thus approaches to her biology, and health, are centered around her capacity to reproduce.[16] And that capacity is thought to involve messy hormones, behavior, and physiology, rendering the female too complicated and a poor subject for general assessment of nonreproductive-related aspects of biology, health, and well-being. Because of this belief, many experiments, or rather experimenters, assume that how something affects a 3G male (be it a mouse or human) is the best proxy for the rest of the species. This approach assumes minimal relevant variation among 3G males and too much variation among 3G-female bodies. This belief has kept female animals, and animals who do not fit neatly into 3G-male categories, out of much laboratory research.[17] For the last century, based on the binary view, most of the medical testing world[18] saw male mice (and male humans) as the ideal baseline, the "normal" for understanding bodies. This commitment to a simple reproductive physiology binary as a key factor when it comes to researching health and medicine is not supported by current understanding of human biological variation and has resulted in medical science doing outsized harm to women.[19]

To be clear, rejecting a binary view of sex biology in medical research and treatment does not mean ignoring patterns of variation across and between bodies, including those associated with uteri, gonads, genes, and genitals.[20] There are critical patterns of physiological variation related to sex biology, and they often matter for issues of health. However, to assume that the most basic dividing line for humans is between two typical sets of reproductive physiology, or that the optimal approach in medical science is to minimize variation in test subjects by avoiding complexity, has negative outcomes.

Take drug reactions. Most reports assert that women (generally meaning 3G females)[21] are about 1.5–2 times as likely as men (generally meaning 3G males) to experience severe side effects when using pharmacological treatments. Women also may not derive the same benefits from many drugs as do men (on average). This is usually assumed to be because of differences in 3G-male and 3G-female biologies. But is that the case?

If one restricts comparisons to 3G categories, most of the documented variation in effects of drugs on 3G males and 3G females is not based solely on body size. There are other physiological processes at play, possibly related to aspects of sex biology. One might take this fact at face value and argue that this reinforces a binary view. It does not. One aspect likely affecting these outcomes is that most of the dosages for drugs were, up until recently, largely based on the clinical testing done solely, or primarily, in men.[22] Another aspect is that the percentage of women's adverse side effects may be overestimated because women (the gender) use prescription drugs more than men (the gender) and the databases are not adjusted for actual drug-use rates by men and women (by 3G-sex category).[23] A third explanation for possible differences in outcomes is that "men" and "women" are not discrete biological categories, so testing pharmaceuticals in the categories "men" and "women" is not effectively assessing the range of human biological variation (the actual people) taking the drugs. Finally, human bodies and their physiologies are shaped by their lived experiences, so gendered cultural facets may affect how bodies and physiologies respond to drugs.

For example, there is a long-standing argument that the sleeping aide zolpidem (Ambien) affects women differently than men. The development of the original dosage data for zolpidem came from testing on men (assumedly 3G males), and because women (here meaning 3G females) are, on average, smaller in overall body and muscle mass than men, the assumption was that the same dose would affect women more than men (on average). Recent testing on women (3G females) shows that they are not more heavily affected by the drug, nor are there different chemical impacts by zolpidem in women's bodies. To the contrary, it appears that there is a slower (up to 35 percent) clearance time of the drug in 3G females (on average) as compared to that in 3G males. This is not due to weight or BMI. Something else is going on. The initial binary-based solution was to blanketly call for reduced dosages in women relative to men, which can result in the drug having less effect and thus failing to achieve its target goal (battling insomnia) in women. But such an approach does not get to why the differences exist. Asking about the actual physiological response, rather than assuming 3G males and 3G females are different kinds of humans, is a better approach. It is likely that zolpidem clearance in humans is mediated, in part, by a system of enzymes called Cytochrome P450 (CYP), and their activity might be related to aspects of testosterone.[24] If this is the case (which is not yet confirmed), then the attention should be focused on the varying levels of acting testosterone in attenuating the effectiveness of zolpidem. Testosterone is not characterizable as a male or female hormone, but rather by variation in circulating levels across humans, with 3G males usually having much higher levels than 3G females.

What appears to be in play in the zolpidem case, and in the majority of assessments of drug-impact variation, is that the sex-binary assumption of males and females as two discrete categories misses the substantial physiological variation in humans within, between, and across sex biologies, gender, and 3G categories.[25] Rather than two categories (men and women) as the comparison, the patterns of variation in enzymatic activity, endocrine function, muscle activity, and so on should be the focus for understanding the specifics of drug action in human systems. One category (males) cannot stand in for all humans. That category, even if restricted to 3G males, in and of itself does not effectively capture all the variation in each of its own constituent Gs (genetic variants, testes variants, and genital variants). For example, there is substantial variation in muscle mass (as much as 100 percent) and massive variation in postpubertal circulating testosterone (by as much as 300 percent) between 3G-male individuals.[26] The solution is also not simply to add an equal number of 3G females to the testing sample (although that is a good start). We know that human females vary substantially even within the 3G definition, in body size,[27] in genital morphology,[28] breast biomechanics,[29] and endocrine dynamics.[30] The binary two-kinds-of-human view restricts effective scientific understanding of pharmacological impacts in humans.

Shifting to psychiatric conditions, many neurobehavioral disorders differ in prevalence between the categories of men and women and boys and girls. Recognized syndromes such as autistic spectrum, ADHD, dyslexia, depression, anxiety, dementia, and eating disorders all have varying frequencies across genders, ethnicities, and nations. Such syndromes are complex and dynamic, and are almost always biocultural. Gender/sex clearly plays a role in many (or all) of these conditions. These conditions represent complex interlacing of physiological, neurological, social, experiential, and individual processes. Although studies of brain variation situated in the sex binary often intend to offer to a better understanding of health disparities in psychiatric and psychological conditions, their assumptions of a "male" and "female" brain are likely obfuscating the actual relationships as the binary frame excludes the complexities of interindividual neurobiological variation and human biocultural experience.[31]

Cardiac Disease

Since 1984, more women than men have died of ischemic heart disease and heart failure each year, and yet more men (here meaning 3G males) have been diagnosed with heart disease than women (here meaning 3G females).[32] Doctors diagnose heart conditions by a combination of clinical presentation, biomarkers, and imaging. Obstructive coronary artery disease (CAD) remains the current focus of most assessment and therapeutic strategies, but women have lower rates of typical CAD than men. Women also have a higher rate of diverse symptoms and adverse outcomes likely linked to coronary and microvascular dysfunction.[33] Also, understanding of antiplatelet therapy, a primary tool of cardiovascular disease prevention, was, until very recently, largely based on research with almost exclusively 3G males, and it appears that there may be a wider range of variation in function of those drugs in 3G females.[34] There is also substantive evidence that the manifestations of cardiac disease and distress, and the effectiveness of various treatments, vary across humans (a lot), and one salient aspect of that variation is in the categories "men" and "women," which are not simply biological. Some of this gender difference is likely linked to variations in 3G-sex-related physiology, but the data are clear that socioeconomic status, nutrition, smoking, alcohol use, racism, sexism, stress hormones, inflammation patterns, immune cell function diversity, and cellular aging all play key roles as well.[35] This pattern suggests that rather than a binary man/woman or 3G-male/3G-female frame, which much of the medical world currently holds,[36] a dynamic gender/sex and biocultural frame should be applied to the issue of cardiac disease and its treatments.

Organ Transplants

While there is no evidence that most organs originate in a binary fashion (3G-male and 3G-female organs) in humans, there are on average worse outcomes for transplants from 3G-female donors regardless of the 3G sex of the recipient. Also, kidneys and hearts (but not livers) transplanted from 3G females have specifically higher failure rates in 3G-male recipients.[37] One might take a binary view and argue that this indicates some level of "female" organs and some kind of incompatibly with "male" immune systems. And one would likely be wrong. It turns out that more 3G females (and women in general) donate organs than do 3G males and men. Those 3G females and women who donate are also generally sicker and older than the population on average and a larger percentage of the 3G males who receive organ transplants are also generally sicker and older than the population on average. This pattern creates a dynamic that is not best examined through only a sex-biology lens. Rather, it's most likely that organs are responsive to effects of gender/sex and other aspects of life. Thus, the details of the lives of the donors and recipients and the traumas—social and physiological—within them are as likely as, if not more, relevant to transplant success than if donor was XX or XY genetically. This situation is described in a large review entitled "How Sex and Gender Affect Transplantation," in which the twenty-six authors state, "In this review, we summarize the data regarding sex- and gender-based disparity in adult and pediatric kidney, liver, lung, heart, and hematopoietic stem cell transplantation and argue that there are not only biological but also psychological and socioeconomic issues that contribute to disparity in the outcome, as well as an inequitable access to transplantation for women and girls."[38]

Pregnancy and Related Physiological Changes

There is a specific set of organs that, due to mammalian reproductive biology, establishes critical patterns of variation between most 3G-female and 3G-male bodies. The dynamics of gestation and lactation, as already noted multiple times in this book, matter immensely and group those who can get pregnant and do get pregnant into a cluster of relevance to medical research and treatment. However, the binary approach generally deployed by the medical world does not cover everything here.

Not all of those who get pregnant and give birth are classified or identify as women, or 3G females, and there are also plenty of individuals who have the physiology for gestation and lactation but cannot or do not get pregnant and give birth. Thus, the focus of research into pregnancy and birth should not simply be on women or 3G females as one of the two categories in a binary. Rather the focus should be on the patterns, dynamics, and experiences of the range of those who can and do get pregnant. For example, in 2020 the National Academies of Science, Engineering, and Medicine, examining the data and relevant patterns of variation changed their terms for research in this arena to "pregnant people" or "pregnant individuals" in place of "pregnant women."[39]

While sex biology plays a core role in pregnancy, a mountain of data supports the assertion that the individual experience of pregnancy, like so much else in human lives, strongly intertwines with social, economic, political, racialized, and related aspects.[40] Even something as specific as where one gives birth can have dramatic impacts on bodies and lives.[41] Pregnancy, like so much of the human, is also biocultural. There is a substantial range of physiological variation between individuals who can and do get pregnant in terms of length of gestation (which varies by as much as many weeks), hypertensive disorders, labor/delivery experience, and much more.[42] Given this range of variation in biology, and the experience and context of pregnancy and birth, rather than seeing women, people who can get pregnant, and people who do get pregnant as a monolithic category, it makes better scientific and social sense to examine the patterns and processes of the variation across and among those who do get pregnant to better facilitate healthy outcomes.

Humans who gestate and give birth undergo a suite of physiological dynamics that are not present in the same form and intensity in other human bodies. Gestating and giving birth play a core role in the construction of gendered perceptions about, and expectations of, women. Many gendered, and medicalized, perspectives present pregnancy as debilitating or as some form of malady. This is common in patriarchal systems that argue for hyperdifferentiation between women's and men's places and capacities in society, often promoting men in the public sphere of power and overemphasizing women as tied to the home and caretaking. But this use of gestation as a tool of restrictions and oppression is not justified biologically. While the impacts of gestation are physiologically substantive and have distinctive effects on bodies, they are not necessarily the restrictive, debilitating constraint many cultural and gendered assumptions assert they are. Pregnant individuals can usually retain their typical levels of social and physical activity and physiological functioning across the pregnancy.[43] And, because the human reproductive system evolved as a cooperative system involving many individuals beyond the mother, the individual gestating and giving birth's physiological system is "expecting" a range of collaborative action by other individuals involved (by their family/community). However, because of the misguided binary-based belief about limitation in social roles and physical incapacitation in humans who gestate and give birth, there is often robust social, economic, and political inequity for all women.

Sex Contextualism in Medicine

Approaching sex biology as a culturally contextualized patterned variation across and within categories is a framework that helps improve medical care by rejecting a simple binary and showing that investing in actual understanding of biological and social variation and patterns across bodies is key to effectively linking biological diversity to facilitating health. Drawing on the conceptual framework of "sex contextualism" can be particularly useful when it comes to medicine and biomedical research.[44] This view emphasizes that "male" and "female" or "men" and "women" do not mean the same things in all contexts, nor are these the only subclasses or categories that are applicable or useful in biomedical research.[45] As there is often substantial variation in some aspects of biology within the categories of "men" and "women," using them as the "bins" everyone gets put into for a given biomedical assessment, as opposed to focusing on the patterns of variation in the biology related to the medical issues of interest, can lead to missed opportunities for achieving health.[46] Think of the zolpidem example and whether or not "male" and "female" are the best categories for comparisons in that case (they aren't), or the concern about biological variability in mice and humans classified as "female" and how this has affected medical research by keeping "females" and their complexity out of it. The data are clear; the medical world should think beyond a sex binary as the only way to ask questions about health. Sex contextualism, by contrast, recognizes the pluralism and context-specificity of operationalizations of "sex" across medical research and urges practitioners to attend to, be clear about, and be consistent with the uses and meanings of the classification used in the design, interpretation, and communication of that research.[47] This approach in medicine may also be important for the sporting world. Sports could also benefit from thinking more deeply about the use, meaning, and impact of sex categories, adding nuance and context to better engage human variation and athletic performance.

Sports and the Definition of Woman

Caster Semenya is a South African athlete, winner of two Olympic gold medals and three World Championships in the women's 800-meter race. After her streak of wins, Ms. Semenya was forced to undergo "sex confirmation" testing, becoming a center of attention in the battle of sex and gender in the sporting world. Her case, and others, resulted in various sets of debates, bans, and restrictions on women athletes. Ms. Semenya, identified as female on her birth certificate, has genitals concordant with that classification, grew up a girl, and is a woman. However, she is not a 3G female. Ms. Semenya has higher-circulating testosterone than is typical for 3G females, a Y chromosome, and internal gonads that are the equivalent of undescended testes.[48] Ms. Semenya is a woman whose sex biology is outside what is typical for 3G females but well within the range of variation found in our species. The binary view, however, does not allow space for her.

Between 2017 and 2019, the International Olympic Committee and World Athletics, the two main groups controlling track events, barred women who had naturally occurring levels of testosterone (T) above 5 nmol/L from competing in races between 400 meters and 1,500 meters (including hurdles). The assertion was that such a level of circulating T was outside of the "normal" range for "women" and that it veered into the range of "men," thus offering an unfair competitive edge to those women. In making this decision, the groups drew on research suggesting that while testosterone levels did not correlate with performance in most sporting events, there were some correlations found between circulating testosterone and levels of success in the 400m, 400m hurdles, and 800m races, and the hammer throw and pole vault for women athletes. Researchers argued that a higher testosterone level seemed to confer a 1.8–2.8 percent competitive advantage, but the actual mechanism of this advantage was not clear. This same study found that male athletes in throwing events had lower than expected testosterone levels and that testosterone did not appear to play a key role in longer-distance running or in most other sporting events assessed for men or women.[49] In a 2017 article, the main group of researchers relied on by the International Olympic Committee and World Athletics concluded that testosterone does not have a biasing effect in most sporting events regardless of the gender classification of the event and that differences in circulating testosterone have no biasing effect in any of the men's categories, but that there is, at least in a few track events, "a significant role of endogenous androgens for athletic performance in women."[50] However, in 2021, after publishing a few more studies arguing that higher testosterone benefits women in some track events, that same group of researchers published a little-noticed correction to their earlier work. In it, they stated that their conclusions about the role of testosterone on athletic performance in women elite athletes "should be interpreted based on an exploratory study setting where it was not possible to adjust for potential confounding factors." And they concluded, "To be explicit, there is no confirmatory evidence for causality in the observed relationships reported."[51] In short, there are patterns of associations between testosterone and many other bodily and experiential factors that influence, and may potentially bias, outcomes in sporting events, but the specific relationships, especially whether it is the level of testosterone that makes the difference, are not clear.[52]

In some, but certainly not all, sporting competitions, upper-body strength, per-unit of area muscle strength, overall body size, and leg length and muscle performance are biological variables that may offer specific benefits. They are also aspects of human bodies that are affected, often substantially, by variation in sex biology. Because these variants do manifest, on average, proportionately differently in 3G females and 3G males, and these 3G categories are most often societally correlated with the categories *men* and *women*, many sports today have developed men's and woman's sports where the competitions are gender-uniform, such as basketball, baseball, football (soccer), tennis, and track and field events.[53] Given recent societal changes (in some places) to accept a broader range of gendered identities, there are increasing situations where individuals who were assigned a gender at birth and have changed that gender (are now transgendered individuals) wish to compete in their current gender category. At the same time, there are also individuals, as with Ms. Semenya, assigned a gender at birth, who remain in that gender, but do not fully fit the current assumptions about the biology "underlying" that gender, who are being excluded from competitions based on a highly specific (and culturally structured) definition of "woman" (basically, how much testosterone the body naturally produces). It is worth noting that being of the right "sex" is brought up as an issue in women's sports but not in men's.

Before making any further comments on these cases, we must recognize the fact that women's professional and amateur sports are relatively new as organizations and institutions, underfunded, and undersupported relative to men's sports. There is abundant evidence that societal gender structures and gender/sex dynamics of girls' and women's lives structure their bodies and possibilities affecting their capacities, performances, and risk of injury in sports differently than those of boys and men who are trained for sports.[54] This gendered training dynamic is illustrated in running sports, where the gender differentials in performance are the smallest at elite levels where the training is at the highest caliber and closer to equity as opposed to larger differences at lower and amateur levels where training and infrastructure quality and options are highly divergent in quality (primarily favoring boys and men).[55] Any discussion of women athletes' capacities and performance must recognize that there is an underlying inequity resulting in women athletes not yet achieving the full potential relative to their physiological capacities at the degree to which men athletes do.

Coming back to the two cases, transwomen and women who are defined out of being a woman by testosterone levels, one notes that each is different both in relation to the dynamics of sex biology and to the ways in which gender/sex works and gender is constructed, but both are united as being identified as a "problem" by the binary view. In the case of defining what a woman *is* by testosterone production, this is clearly not simply a biological measure, as it reflects a social construct of what the "correct" range of testosterone should be. While testosterone and other androgens often play roles in much physiological functioning related to physical activity, it is not clear if or how an individual's physical performance at sports, including limb length, training regime, muscle density, oxygen-carrying capacity of the red blood cells, lung capacity, and so on, translates into a deciding factor as to one's gender and broadscale physiological capacity. The decision to define who is a woman based on specific ranges of testosterone levels is a social and cultural one, not a biological one, and it has a range of ethical and societal implications.[56]

For the case of transwomen competing in women's sporting events, it is a bit more complicated. If an individual is competing in a sport where aspects of sex biology such as upper-body strength, muscle density, or overall body size offer benefits, then an individual may indeed have an advantage *if* they are at the top end of the distribution of variation in those bodily areas relative to their competitors. But the structure and patterns of such advantage vary by sport, individual life histories, body variation, and other variables.[57] This is the case already in gender-uniform sports, as certain individuals have biological variants that offer them a particular set of advantages (think of the swimmer Michael Phelps, a 3G male and a man, relative to the majority of men he competed against[58]). However, in the case of transwomen, it is argued that some of those benefits will be substantially outsized because of lingering effects of their pretransition physiology. But this assumes that the individual transitioned postpuberty from a body with testes *and* that they are in the upper end of 3G-male physiological outcomes as opposed to being in the large area of overlap between 3G-male and 3G-female bodies. Given the extremely small number of transgendered athletes overall and the even smaller number of those individuals who are winning an outsized percentage of their competitions, it is difficult, problematic, and largely unscientific to make any broad-scale blanket generalizations of this situation. And any rules created from such generalizations, especially if based exclusively on natal gonads and timing of transition, ignore the substantive variation in bodies and capacities (which is unfortunately what World Athletics did[59]). However, one thing is clear: the current gender inclusion and exclusion system for many sporting events, leagues, and competitions is insufficient to engage, effectively, the sex-biology variation and the gender/sex reality of humans.[60]

This focus on sports events, athletes' bodies, and testosterone is interesting and matters for the athletes and the sporting world. But it is a poor venue to examine, and think about, variation in sex biology in humans. Only a minuscule percentage of humans have the pattern of biological variation that confers bodily capacity, skillset, and opportunity to be elite athletes or even athletes in amateur and school competitive contexts. There are about 85,000–95,000 professional athletes *in the entire world* and about 170,000 individuals worldwide who compete for Olympic slots.[61] To put this in perspective: there are only between 6,000 and 11,000 professional athletes in the entire United States, out of a population of over 330,000,000. This means that elite athletes make up .0012 percent of living humans globally and .003 percent of people in the United States. Using one thousandth of one percent of humans, who are mostly at the extreme ends of the human range for many biological factors and with a lifetime training related to their sports, as the key group to understand human-wide patterns of sex biology is misleading and faulty science. The focus on athletes to legislate issues of transgender makes even less sense when one considers that only 0.6 percent of the population 13 years of age and above in the United States identifies as transgendered, and only a tiny percentage of them are athletes.[62]

In the day-to-day life of the vast majority of humans, a slight competitive edge in running an 800-meter race is probably not relevant. Variation in circulating levels of testosterone is not the sole determinant, or necessarily a main determinant, of any individual's behavior, capacities, experience, or identity.[63] This is not to state that testosterone is not an important hormone for humans (and most animals), because it is, as are estrogen, progesterone, and so on. It is also not to state that T is not associated with many aspects of physiology and has some differing impacts on bodily systems across humans, specifically in individuals with testes or individuals who receive external testosterone. Testosterone has a range of important functions and impacts in the human body in multiple manners. However, if we are truly interested in an accurate and comprehensive understanding of sex biology, then focusing on something as recent and hyperculturally modified at professional/elite athletics is a poor place to look.

Restrooms

In 2016, the US state of North Carolina passed a law prohibiting transgender individuals from using the restroom that corresponded to their gender. This action was part of the political, legal, and public debates over whether transgender individuals should be able to use public restrooms (including locker rooms and changing rooms) that match their gender or if they should be forced to use ones that match a social categorization of the person based on which genitals they have (or had) and how they were assigned at birth. Between 2021 and 2023, legislatures in 34 US states introduced over 300 anti-transgender bills reflecting a wide range of proposed restrictions on issues of health, legal rights, and access for individuals who did not fit certain beliefs about gender and sex biology.[64] By early 2024, there were more than 450 active anti-transgender bills moving through the legislatures in 41 states.[65] In June 2023, the state of Florida implemented House Bill 1521, which mandates that transgender individuals must use public restrooms that correspond with the sex assigned to them at birth, even if they have legally updated their gender on their birth certificate and driver's license. If they do not comply, they can be charged with trespassing—punishable by up to a year in jail. In January of 2024, Utah followed with a similar law. These laws and their associated outcomes are a clear, powerful, and targeted attempt, based on the binary notion of sex biology and gender, to curtail the rights of a group of humans whose bodies and lives challenge the binary view.

Everyday people encounter the world in a myriad of different ways because of the combinations of their distinct sex-biology variations, their gender/sex, and a society's expectations of gender. In some societies, restrooms in public, in school, and in the workplace are divided along a binary gender system with separate facilities for men/boys and women/girls. For example, in the United States, depending on one's gender classification, the structures and locations for urinating and defecating (urinals, toilets, stalls, dividers, etc.) are quite different as are the layout and often the sanitary conditions of the restrooms. Some of the differences are asserted to be connected to genital differences and the capacity to urinate while standing. But most of the structures are highly gendered and reflect not so much the physical necessities of sex-biology variation (all humans can effectively urinate and defecate when sitting) but the cultural assumptions about what that variation means. It is also an explicit cultural assertion that a binary division is the only acceptable option for human gender/sex regarding urination, defecation, and washing up. Those individuals with sex-biology and gender/sex variation fully within the range of the human pattern but not fitting this specific set of genital-based (and broadly 3G) binary assumptions must then decide which restroom to enter and use and risk, in many cases, harassment, violence, or legal repercussions if someone else decides they've chosen the "wrong" one.[66]

The restroom example is just one of so many public misperceptions rooted in the binary. Whether it is in the workplace, the political or religious sphere, school, or broader public venues, everyday people vary a lot, but most assumptions about what that variation is and what it means are highly restrictive and ignore the actual variation in the human system. Simplistic binary assertions about how men or women "are" and "should" be, and assumptions that everyone with the same type of genitals or gonads shares the same set of ways of being, permeate the world, making it difficult not only for those who do not fit the general assumptions for the categories of men and women, but also those who do.

The Binary Is Wrong and Harmful

This book has endeavored to introduce what is actually known about sex biology and the human experience and to explain why the most accurate way to discuss and research it is via a biocultural framework. The best science requires a full consideration of patterns and diversity and the recognition that humans have gender/sex. Thinking broadly about the evolution of sex, sex biology across the animal kingdom, and the ways in which our biological understanding of sex has changed over time helps us see diversity in sex biology as typical. There are few simple one-to-one universal "truths" about being a female or male, and treating those two categories as different kinds of being is not supported by the science of sex biology. Reviewing human evolutionary history and the patterns of sex biology in current human bodies opens us up to a better understanding of why and how humans are in the world and rejects simplistic binary assertions about men and women.

The take-home message is that patterns of sex biology, while always important, are not the exclusive determinants of the entirety of or the specifics of any given individual's gender/sex experience or their experience of self. The biological, behavioral, and social variation exhibited by humans is almost always substantively broader, and more dynamic, than culturally created binary descriptions and boundaries for gender and sex. Moving past the binary view does not mean ignoring variation and differences across humanity, including those related to sex biology. Rather, the necessity to develop better understandings of those differences, and that variation, is exactly why there is a need to move past the sex and gender binary. Let's do everything we can to make that happen.

NOTES

1. The Evolution of Sex

1. The earliest life probably started with RNA (single-stranded genetic material), and then DNA (double-stranded genetic material) emerged a bit later

2. The bacteria are divided into archaebacteria and eubacteria, but much current work separates the two further into archaea and bacteria as distinct evolutionary lineages

3. U. Goodenough and J. Heitman. 2014. "Origins of Eukaryotic Sexual Reproduction." *Cold Spring Harbor Perspectives in Biology* 6 (3): a016154. https://doi.org/10.1101/cshperspect.a016154; D. Bachtrog, J. E. Mank, C. L. Peichel, M. Kirkpatrick, S. P. Otto, T. L. Ashman, M. W. Hahn, et al. 2014. "Tree of Sex Consortium. Sex Determination: Why So Many Ways of Doing It?" *PLoS Biology* 12 (7): e1001899. https://doi.org/10.1371/journal.pbio.1001899.

4. G. C. Williams. 1975. *Sex and Evolution.* Monographs in Population Biology 8. Princeton, NJ: Princeton University Press, v.

5. For a very basic overview, see C. Zimmer. 2009. "On the Origin of Sexual Reproduction." *Science* 324 (5932): https//doi.org/10.1126/science.324_1254. For two much more in-depth, classic overviews, see G. C. Williams. 1975. *Sex and Evolution.* Monographs in Population Biology 8. Princeton, NJ: Princeton University Press; and U. W. Goodenough. 1985. "Origins and Evolution of Eukaryotic Sex." In *Origins and Evolution of Sex*, edited by H. O. Halvorson and A. Monroy, 123–40. New York: Alan R. Liss. For more recent reviews, see U. Goodenough and J. Heitman. 2014. "Origins of Eukaryotic Sexual Reproduction." *Cold Spring Harbor Perspectives in Biology* 6 (3): a016154. https//doi.org/10.1101/cshperspect.a016154.

6. J. Lehtonen and G. A. Parker. 2014. "Gamete Competition, Gamete Limitation, and the Evolution of the Two Sexes." *Molecular Human Reproduction* 20 (12): 1161–68. https://doi.org/10.1093/molehr/gau068; D. Bachtrog, J. E. Mank, C. L. Peichel, M. Kirkpatrick, S. P. Otto, T. L. Ashman, M. W. Hahn, et al. 2014. "Tree of Sex Consortium. Sex Determination: Why So Many Ways of Doing It?" *PLoS Biology* 12 (7): e1001899. https://doi.org/10.1371/journal.pbio.1001899.

7. Since this book is really just focused on the animals, we won't even try to get into the sex biology of plants, which is far more complicated. If you are interested, read D. Charlesworth. 2002. "Plant Sex Determination and Sex Chromosomes." *Heredity* 88: 94–101. https://doi.org/10.1038/sj.hdy.6800016; J. R. Pannell. 2017. "Plant Sex Determination." *Current Biology* 27 (5): R191–R197. https://.doi.org/10.1016/j.cub.2017.01.052; and especially B. Subramaniam and M. Bartlett. 2023. "Re-imagining Reproduction: The Queer Possibilities of Plants." *Integrative and Comparative Biology* 63 (4): 946–59. https://doi.org/10.1093/icb/icad012.

8. T. Laqueur. 1990. *Making Sex: Body and Gender from the Greeks to Freud*. Cambridge, MA: Harvard University Press.

9. V. Sanz. 2017. "No Way Out of the Binary: A Critical History of the Scientific Production of Sex." *Signs: Journal of Women in Culture and Society* 43 (1): 1–27. https://doi.org/10.1086/692517; S. Richardson. 2013. *Sex Itself: The Search for Male and Female in the Human Genome*. Chicago: The University of Chicago Press; see also E. Martin. 1991. "The Egg and the Sperm: How Science Has Constructed a Romance Based on Stereotypical Male-Female Roles." *Signs* 16 (3): 485–501.

10. C. Darwin. 1871. *The Descent of Man and Selection in Relation to Sex*. London: John Murry.

11. A. J. Bateman. 1948. "Intrasexual Selection in Drosophila." *Heredity* 2:349–68.

12. Bateman's assertions were that females are limited in how many gametes they can produce due to the size and energetic costs of ova, and thus their investment in each gamete is very high. Basically, Bateman (working with fruit flies) argued that males can produce unlimited amounts of small sperm, while females can only produce a limited number of large ova. So, given this difference in costs to the organisms, males should try to mate with as many females as possible (get as much sperm out there as possible), but females, with costly, limited ova, should be coy, sexually passive, and very choosy. This is what Darwin said in 1871 about animal evolution in general (including humans), but in 1948, Angus Bateman put the cause of this assumed reality squarely on the fact of anisogamy.

13. G. C. Williams. 1966. *Adaptation and Natural Selection*. Princeton, NJ: Princeton University Press; R. Trivers. 1972. "Parental Investment and Sexual Selection." In *Sexual Selection and the Descent of Man*, edited by B. Campbell, 52–95. Chicago: Aldine.

14. E. O. Wilson. 1975. *Sociobiology: The New Synthesis*. Cambridge, MA: Belknap Press of Harvard University Press; G. A. Parker. 1979. "Sexual Selection and Sexual Conflict." In *Sexual Selection and Reproductive Competition*, edited by M. S. Blum and N. A. Blum, 123–66. New York: Academic Press. Please note all of the most prominent scholars making these arguments are men: Darwin, Bateman, R. L. Trivers, E. O. Wilson, and G. A. Parker.

15. Z. Tang-Martinez and T. B. Ryder. 2005. "The Problem with Paradigms: Bateman's Worldview as a Case Study." *Integrative and Comparative Biology* 45: 821–30.

See also P. Gowaty, Y. Kim, and W. Anderson. 2012. "No Evidence of Sexual Selection in a Repetition of Bateman's Classic Study of *Drosophila melanogaster*." *Proceedings of the National Academy of Sciences of the United States of America* 109: 11740–45. https://doi.org/10.1073/pnas.1207851109; H. Kokko and M. Jennions. 2003. "It Takes Two to Tango." *Trends in Ecology and Evolution* 18 (3): 103–4; C. M. Drea. 2005. "Bateman Revisited: The Reproductive Tactics of Female Primates." *Integrative and Comparative Biology* 45 (5): 915–23. https://doi.org/10.1093/icb/45.5.915; M. Borgerhoff-Mulder. 2004. "Are Men and Women Really So Different?" *Trends in Ecology and Evolution* 19 (1): 3–6.

16. M. Ah-King. 2013. "On Anisogamy and the Evolution of 'Sex Roles.'" *Trends in Ecology and Evolution* 28 (1): 1–2. https://doi.org/10.1016/j.tree.2012.04.004. A. Fausto-Sterling. 2020. *Sexing the Body: Gender Politics and the Construction of Sexuality*. London: Hachette; L. Z. DuBois and H. Shattuck-Heidorn. 2021. "Challenging the Binary: Gender/Sex and the Bio-logics of Normalcy." *American Journal of Human Biology* 33 (5): e23623. https://doi.org/10.1002/ajhb.23623; J. F. McLaughlin, Kinsey M. Brock, I. Gates, P. Anisha, M. Piattoni, A. Rossi, and S. E. Lipshutz. 2023. "Multivariate Models of Animal Sex: Breaking Binaries Leads to a Better Understanding of Ecology and Evolution." *Integrative and Comparative Biology* 63 (4): 891–906. https://doi.org/10.1093/icb/icad027; C. T. Ross, P. L. Hooper, J. E. Smith, A. V. Jaeggi, E. A. Smith, S. Gavrilets, F. T. Zohora, et al. 2023. "Reproductive Inequality in Humans and Other Mammals." *Proceedings of the National Academy of Sciences of the United States of America* 120 (22): e2220124120. https://doi.org/10.1073/pnas.2220124120.

17. Much of the biological literature on nonhuman animals still uses the term "hermaphrodite" for individuals who have both gonad types (or an ovotestis). "Hermaphrodite" is no longer used for humans, and increasingly less so for other animals, due to historical baggage and the fact that the term "intersex" is more expansive and encompasses multiple biological variants regarding gonads and other sex biology.

18. Except not always. There are a number of fish and reptiles that make these structures but still store them internally until the zygotes develop into juvenile organisms and are capable of making it on their own. Then they excrete them.

19. Many egg-laying species also have genitals for assisting in the internal fusion of gametes.

20. For relatively concise and pretty good overviews of animal reproductive systems, see G. C. Kent. 2021. "Animal Reproductive System." *Encyclopedia Britannica*, November 19. https://www.britannica.com/science/animal-reproductive-system.

21. This system is a bit different in marsupial and placental mammals. In placentals, all the fetal development is internal, but in marsupials the last portions of it take place in a sack external to the reproductive tract.

22. Z. M. Thayer, J. Rutherford, and C. W. Kuzawa. 2020. "The Maternal Nutritional Buffering Model: An Evolutionary Framework for Pregnancy Nutritional Intervention." *Evolution, Medicine, and Public Health* 1: 14–27.

23. See L. Cooke. 2022. *Bitch: On the Females of the Species*. New York: Basic Books.

24. See, e.g., A. Liker, R. P. Freckleton, V. Remeš, and T. Székely. 2015. "Sex Differences in Parental Care: Gametic Investment, Sexual Selection, and Social Environment." *Evolution* 69 (11): 2862–75; H. Kokko and M. D. Jennions. 2008. "Parental Investment, Sexual Selection and Sex Ratios." *Journal of Evolutionary Biology* 21 (4): 919–48; Drea, "Bateman Revisited."

25. R. L. Goldberg, P. A. Downing, A. S. Griffin, and J. P. Green. 2020. "The Costs and Benefits of Paternal Care in Fish: A Meta-Analysis." *Proceedings of the Royal Society B* 287 (1935): https://doi.org/10.1098/rspb.2020.1759.

26. A. I. Furness, I. Capellini. 2019. "The Evolution of Parental Care Diversity in Amphibians." *Nature Communications* 10: 4709. https://doi.org/10.1038/s41467-019-12608-5.

27. A. Cockburn. 2006. "Prevalence of Different Modes of Parental Care in Birds." *Proceedings of the Royal Society: Biology* 273 (1592): 1375–83.

28. See J. K. Rilling and L. J. Young. 2014. "The Biology of Mammalian Parenting and Its Effect on Offspring Social Development." *Science* 345 (6198): 771–76; L. S. B. Hrdy. 2009. *Mother and Others: The Evolutionary Origins of Mutual Understanding*. New York: Belknap; L. T. Gettler. 2016. "Becoming DADS: Considering the Role of Cultural Context and Developmental Plasticity for Paternal Socioendocrinology." *Current Anthropology* 57: S38–S51; S. Rosenbaum and L. T. Gettler. 2018. "With a Little Help from Her Friends (and Family) Part I: The Ecology and Evolution of Non-maternal Care in Mammals." *Physiology and Behavior* 193: 1–11; and S. Rosenbaum and L. T. Gettler. 2018. "With a Little Help from Her Friends (and Family) Part II: Non-maternal Caregiving and Physiology Caregiving in Mammals." *Physiology and Behavior* 193: 12–24.

29. This is called "niche construction," and it is a very salient and important part of evolutionary processes. In particular, patterns of niche construction can play central roles in expanding and restricting the roles and types of variation possible in sex biology. This will be a central theme in chapters 2 and 3. For a good overview article, see here: K. Laland, B. Matthews, and M. W. Feldman. 2016. "An Introduction to Niche Construction Theory." *Evolutionary Ecology* 30: 191–202. And for an online review see here: K. Laland and L. Chiu. 2020. Niche Construction. https://nicheconstruction.com/information/.

30. J. F. McLaughlin, K. M. Brock, I. Gates, A. Pethkar, M. Piattoni, A. Rossi, and S. E. Lipshutz. 2023. "Multivariate Models of Animal Sex: Breaking Binaries Leads to a Better Understanding of Ecology and Evolution." *Integrative and Comparative Biology* 63 (4): 891–906. https://doi.org/10.1093/icb/icad027.

2. Animal Sex Biology: Mixing It Up

1. An analysis of more than four hundred mammalian species indicates that although males tend to be larger than females when dimorphism occurs, males are not larger in most mammals, and suggests a need to revisit other assumptions in sexual selection and sex-biology research. K. J. Tombak, S. B. S. W. Hex, D. I. Rubenstein. 2023. "New Estimates Indicate That Males Are Not Larger Than Females in Most Mammals." *Nature Communications* 15 (1): 1872. https://doi.org/10.1038/s41467-024-45739-5.

2. L. Cooke. 2022. *Bitch: On the Females of the Species*. New York, Basic Books.

3. T. Janicke, I. K. Häderer, M. J. Lajeunesse, and N. Anthes. 2016. "Darwinian Sex Roles Confirmed across the Animal Kingdom." *Science Advances* 2 (2): e1500983; C. M. Drea. 2005. "Bateman Revisited: The Reproductive Tactics of Female Primates." *Integrative and Comparative Biology* 45 (5): 915–23. https://doi.org/10.1093/icb/45.5.915; and Cooke, *Bitch*.

4. Cooke, *Bitch*.

5. There are some species of bee where "workers" do end up laying ova and thus reproducing some haploid offspring, contributing up to 10 percent of the hives' males. D. A. Alves, V. L. Imperatriz-Fonseca, T. M. Francoy, P. S. Santos-Filho, J. Billen, and T. Wenseleers. 2009. "The Queen Is Dead—Long Live the Workers: Intraspecific Parasitism by Workers in the Stingless Bee *Melipona scutellaris*." *Molecular Ecology* 18 (19): 4102–11. https://doi.org/10.1111/j.1365-294X.2009.04323.x.

6. H. Blackmon, L. Ross, D. Bachtrog. 2017. "Sex Determination, Sex Chromosomes, and Karyotype Evolution in Insects." *Journal of Heredity* 108 (1): 78–93.

7. D. J. Diaz Cosin, M. Novo, and R. Fernández. 2010. "Reproduction of Earthworms: Sexual Selection and Parthenogenesis." In *Biology of Earthworms*, edited by A. Karaca, 69–86. Berlin: Springer-Verlag.

8. Multiple sets of chromosomes, not just two.

9. C. H. Chandler, G. E. Chadderdon, P. C. Phillips, I. Dworkin, and F. J. Janzen. 2011. "Experimental Evolution of the *Caenorhabditis elegans* Sex Determination Pathway." *Evolution* 66 (1): 82–93. https://doi.org/10.1111/j.1558-5646.2011.01420.x.

10. E. V. Todd, H. Liu, S. Muncaster, and N. J. Gemmell. 2016. "Bending Genders: The Biology of Natural Sex Change in Fish." *Sexual Development* 10 (5–6): 223–41. https://doi.org/10.1159/000449297.

11. Todd et al., "Bending Genders."

12. K. Semsar and J. Godwin. 2004. "Multiple Mechanisms of Phenotype Development in the Bluehead Wrasse." *Hormones and Behavior* 45: 345–53.

13. B. Oskon. 2014. "For Sex-Changing Fish, Size Matters in Urchin Battle." LiveScience, December 12. https://www.livescience.com/49103-sheephead-fish-recovery-overfishing.html.

14. Todd et al., "Bending Genders."

15. P. L. Munday, T. Kuwamura, and F. J. Kroon. 2010. *Bi-Directional Sex Change in Marine Fishes*. Berkeley: University of California Press.

16. K. T. Moeller. 2013. "Temperature-Dependent Sex Determination in Reptiles." Embryo Project Encyclopedia. https://hdl.handle.net/10776/4214.

17. R. Cox, M. Butler, and H. John-Alder. 2007. "The Evolution of Sexual Size Dimorphism in Reptiles." In *Sex, Size and Gender Roles: Evolutionary Studies of Sexual Size Dimorphism*, edited by D. J. Fairbairn, W. U. Blanckenhorn, and T. Székely. Oxford Academic. https://doi.org/10.1093/acprof:oso/9780199208784.003.0005.

18. K. Harmon. 2010. "No Sex Needed: All-Female Lizard Species Cross Their Chromosomes to Make Babies." *Scientific American*, February 21. https://www.scientificamerican.com/article/asexual-lizards/.

19. A. Lutes, W. Neaves, D. Baumann, et al. 2010. "Sister Chromosome Pairing Maintains Heterozygosity in Parthenogenetic Lizards." *Nature* 464: 283–86. https://doi.org/10.1038/nature08818.

20. R. Somaweera, M. Brien, and R. Shine. 2013. *The Role of Predation in Shaping Crocodilian Natural History*. Herpetological Monographs 27 (1): 23–51.

21. T. Wesołowski. 2004. "The Origin of Parental Care in Birds: A Reassessment." *Behavioral Ecology* 15 (3): 520–23. https://doi.org/10.1093/beheco/arh039.

22. V. Remeš, R. P. Freckleton, J. Tökölyi, A. Liker, and T. Székely. 2015. "The Evolution of Parental Cooperation in Birds." *Proceedings of the National Academy of Sciences of the United States of America* 112 (44): 13603–8. https://doi.org/10.1073/pnas.1512599112.

23. I. P. F. Owens and I. R. Hartley. 1998. "Sexual Dimorphism in Birds: Why Are There So Many Different Forms of Dimorphism?" *Proceedings of the Royal Society: Biology* 265 (1394): 397–407. https://doi.org/10.1098/rspb.1998.0308.

24. W. Webb, D. Brunton, J. Aguirre, D. Thomas, M. Valcu, and J. Dale. 2016. "Female Song Occurs in Songbirds with More Elaborate Female Coloration and Reduced Sexual Dichromatism." *Frontiers in Ecology and Evolution* 4. https://doi.org/10.3389/fevo.2016.00022.

25. C. Arnold. 2016. "The Sparrow with Four Sexes." *Nature* 539: 482–84. https://doi.org/10.1038/539482a; D. J. Newhouse, P. Minx, M. Stager, A. Betuel, Z. A. Cheviron, W. C. Warren, R. A. Gonser, and C. N. Balakrishnan. 2016. "Divergence and Functional Degradation of a Sex Chromosome-Like Supergene." *Current Biology* 26 (3): 344–50. https://doi.org/10.1016/j.cub.2015.11.069.

26. Given the range variations in development, it is physiologically possible that one could have a mix of reproductive physiology that does not fit neatly into typical patterns and be capable of gestation. However, this is likely to be quite rare. See, for instance, B. A. H. Schultz, S. Roberts, A. Rodgers, and K. Ataya. 2009. "Pregnancy

in True Hermaphrodites and All Male Offspring to Date." *Obstetric Gynecology* 113: 534–36. Also, there are a few species of bats where small-gamete producers lactate.

27. Cooke, *Bitch*; Drea, "Bateman Revisited"; S. R. Zajitschek, F. Zajitschek, R. Bonduriansky, R. C. Brooks, W. Cornwell, D. S. Falster, M. Lagisz, et al. 2020. "Sexual Dimorphism in Trait Variability and Its Eco-evolutionary and Statistical Implications." *Elife* (9): e63170. https://doi.org/10.7554/eLife.63170.

28. See C. Delle-Amores. 2024. "Love Them or Hate Them, Hyenas Are Getting the Last Laugh." *National Geographic Magazine*, February 8. https://www.nationalgeographic.com/animals/article/spotted-hyenas-queens.

29. S. E. Glickman, L. G. Frank, J. M. Davidson, E. R. Smith, and P. K. Siiteri. 1987. "Androstenedione May Organize or Activate Sex-Reversed Traits in Female Spotted Hyenas." *Proceedings of the National Academy of Sciences of the United States of America* 84 (10): 3444–47. https://doi.org/10.1073/pnas.84.10.3444.

30. A. Conley, N. J. Place, E. L. Legacki, G. L. Hammond, G. R. Cunha, C. M. Drea, M. L. Weldele, et al. 2020. "Spotted Hyaenas and the Sexual Spectrum: Reproductive Endocrinology and Development." *Journal of Endocrinology* 247 (1): R27–R44. https://doi.org/10.1530/JOE-20-0252.

31. Conley et al., "Spotted Hyaenas and the Sexual Spectrum."

32. And, in fact, there is a solid argument that "monogamy" is not one thing. There can be social monogamy (but having sex with more than one partner) and/or sexual monogamy (sex with only one partner). See A. Fuentes. 1999. "Re-evaluating Primate Monogamy." *American Anthropologist* 100 (4): 890–907; A. Fuentes. 2002. "Patterns and Trends in Primate Pair Bonds." *International Journal of Primatology* 23 (4): 953–78; U. Reichard. 2017. "Monogamy." In *The International Encyclopedia of Primatology*, edited by A. Fuentes, 831–35. London: John Wiley & Sons; E. Fernandez-Duque, M. Huck, S. Van Belle, and A. Di Fiore. 2020. "The Evolution of Pair-Living, Sexual Monogamy, and Cooperative Infant Care: Insights from Research on Wild Owl Monkeys, Titis, Sakis, and Tamarins." *American Journal of Physical Anthropology* 171 (suppl. 70): 118–73. https://doi.org/10.1002/ajpa.24017.

33. M. M. Holmes and B. D. Goldman. 2021. "Social Behavior in Naked Mole-Rats: Individual Differences in Phenotype and Proximate Mechanisms of Mammalian Eusociality." *Advances in Experimental Medicine and Biology* 1319: 35–58. https://doi.org/10.1007/978-3-030-65943-1_2.

34. S. E. Glickman, R. V. Short, and M. B. Renfree. 2005. "Sexual Differentiation in Three Unconventional Mammals: Spotted Hyenas, Elephants and Tammar Wallabies." *Hormones and Behavior* 48: 403–17. https://doi.org/10.1016/j.yhbeh.2005.07.013; Cooke, *Bitch*; Drea, "Bateman Revisited"; Zajitschek et al., "Sexual Dimorphism in Trait Variability."

35. Cooke, *Bitch*.

3. Humans Are Messy

1. M. Cintas-Peña, M. Luciañez-Triviño, R. Montero Artús, A. Bileck, P. Bortel, F. Kanz, K. Rebay-Salisbury, et al. 2023. "Amelogenin Peptide Analyses Reveal Female Leadership in Copper Age Iberia (c. 2900–2650 BC)." *Scientific Reports* 13 (1): 9594. https://doi.org/10.1038/s41598-023-36368-x. But also see later in the book (chapter 5) to recognize that XX or XY are not always accurate indicators of biological sex.

2. T. Laqueur. 1990. *Making Sex: Body and Gender from the Greeks to Freud*. Cambridge, MA: Harvard University Press; V. Sanz. 2017. "No Way Out of the Binary: A Critical History of the Scientific Production of Sex." *Signs: Journal of Women in Culture and Society* 43 (1): 1–27. https://doi.org/10.1086/692517; S. Richardson. 2013. *Sex Itself: The Search for Male and Female in the Human Genome*. Chicago: The University of Chicago Press; E. Martin. 1991. "The Egg and the Sperm: How Science Has Constructed a Romance Based on Stereotypical Male-Female Roles." *Signs* 16 (3): 485–501. https://www.jstor.org/stable/3174586.

3. J. F. McLaughlin, K. M. Brock, I. Gates, A. Pethkar, M. Piattoni, A. Rossi, S. E. Lipshutz. 2023. "Multimodal Models of Animal Sex: Breaking Binaries Leads to a Better Understanding of Ecology and Evolution." *Integrative and Comparative Biology* 63 (4): 891. https://doi.org/10.1093/icb/icad027; see also K. O. Smiley, K. M. Munley, K. Aghi, S. E. Lipshutz, T. M. Patton, D. S. Pradhan, T. K. Solomon-Lane, et al. 2024. "Sex Diversity in the 21st Century: Concepts, Frameworks, and Approaches for the Future of Neuroendocrinology." *Hormones and Behavior* 157: 105445. https://doi.org/10.1016/j.yhbeh.2023.105445.

4. A. Fuentes. 2022. *Race, Monogamy and Other Lies They Told You: Busting Myths about Human Nature*. 2nd ed. Oakland: University of California Press.

5. C. Kuzawa, H. T. Chugani, L. I. Grossman, L. Lipovich, O. Muzik, P. Hof, R. Patrick, D. E.Wildman, et al. 2014. "Metabolic Costs and Evolutionary Implications of Human Brain Development." *Proceedings of the National Academy of Sciences of the United States of America* 111 (36): 13010–15; G. Downey and D. H. Lende. 2012. "Evolution and the Brain." In *The Encultured Brain: An Introduction to Neuroanthropology*, edited by D. H. Lende and G. Downey, 103–38. Cambridge, MA: MIT Press.

6. S. Han and Y. Ma. 2015. "A Culture-Behavior-Brain Loop Model of Human Development." *Trends in Cognitive Science* 19: 666–76; S. Han. 2017. *The Sociocultural Brain*. Oxford: Oxford University Press; C. C. Sherwood and A. Gomez-Robles. 2017. "Brain Plasticity and Human Evolution." *Annual Reviews of Anthropology* 46: 399–419; A. Fuentes. 2019. *Why We Believe: Evolution and the Human Way of Being*. New Haven, CT: Yale University Press/Templeton Press.

7. M. Lock. 2015. "Comprehending the Body in the Era of the Epigenome." *Current Anthropology* 56 (2): 151–77; A. Fuentes. 2023. "Humans Are Biocultural, Science

Should Be Too." *Science* 382: 6672. https://doi.org/10.1126/science.adl151; A. Fuentes and P. Wiessner. 2016. "Reintegrating Anthropology: From Inside Out." *Current Anthropology* 57 (suppl. 13): 3–12; Downey and Lende, "Evolution and the Brain"; T. Leatherman and A. H. Goodman. 2020. "Building on the Biocultural Syntheses: 20 Years and Still Expanding." *American Journal of Human Biology* 32 (4): e23360; T. Ingold and G. Paalson. 2013. *Biosocial Becomings: Integrating Social and Biological Anthropology*. Cambridge: Cambridge University Press.

8. Gender also includes individual expressions of identity expectations and experiences. S. Nanda. 2000. *Gender Diversity: Cross-Cultural Variations*. Long Grove, IL: Waveland Press; J. Lorber. 2022. *The New Gender Paradox*. Oxford: Polity Press.

9. S. de Beauvoir. [1949] 2011. *The Second Sex*. Translated by J. Thurman. New York: Vintage Books, 300.

10. Lorber, *New Gender Paradox*; J. S. Hyde, R. S. Bigler, D. Joel, C. C. Tate, and S. M. van Anders. 2019. "The Future of Sex and Gender in Psychology: Five Challenges to the Gender Binary." *American Psychologist* 74 (2): 171–93. https://doi.org/10.1037/amp0000307; S. M. van Anders. 2022. "Gender/Sex/ual Diversity and Biobehavioral Research." *Psychology of Sexual Orientation and Gender Diversity*, November 10. Advance online publication. https://dx.doi.org/10.1037/sgd0000609.

11. K. Cheslack-Postava and R. M. Jordan-Young. 2012. "Autism Spectrum Disorders: Toward a Gendered Embodiment Model." *Social Sciences and Medicine* 74: 1667–74.

12. A. Zugman, L. M. Alliende, V. Medel, R. A. I. Bethlehem, J. Seidlitz, G. Ringlein, C. Arango, et al. 2023. "Country-Level Gender Inequality Is Associated with Structural Differences in the Brains of Women and Men." *Proceedings of the National Academy of Sciences of the United States of America* 120 (20): e2218782120. https://doi.org/10.1073/pnas.2218782120.

13. This is a recent term and was not used by the societies I reference here, as there were gendered roles beyond the binary male/female in many of them. Generally, the term today includes those individuals whose gender identity does not correspond with the one assigned to them at birth by their culture, usually based on the assessment of genitals at birth, and who seek to live in the other gender role.

14. R. Smith. 2021. "Imperial Terroir: Toward a Queer Molecular Ecology of Colonial Masculinities." *Current Anthropology* 62 (suppl. 23): S155–S168; K. Tallbear. 2015. "Settler Love Is Breaking My Heart." *The Critical Polyamorist Blog: Polyamory, Indigeneity, and Cultural Politics*, May 29. http://www.criticalpolyamorist.com/homeblog/settler-love-is-breaking-my-heart; Q. McLamore. 2023. "Disarming Transphobia: 'Rapid-Onset Gender Dysphoria' Is a Popular Weapon in the Anti-Trans Arsenal. It Is Nothing but Unscientific Bunk." *Aeon*, January 23. https://aeon.co/essays/the-real-reason-why-theres-a-global-rise-in-trans-youth.

15. Nanda, *Gender Diversity*. See also Hyde et al., "Future of Sex and Gender in Psychology."

16. This is especially prominent in gender system of primarily patriarchal societies. A. Saini. 2023. *The Patriarchs: The Origins of Inequality*. New York: Penguin Press.

17. B. E. deMayo, A. E. Jordan, and K. R. Olson. 2022. "Gender Development in Gender Diverse Children." *Annual Review of Developmental Psychology* 4 (1): 207–29; Fuentes, *Race, Monogamy and Other Lies*; A. Fausto-Sterling. 2020. *Sexing the Body: Gender Politics and the Construction of Sexuality*. London: Hachette; C. Fine. 2010. *Delusions of Gender: How Our Minds, Society, and Neurosexism Create Difference*. New York: W. W Norton.

18. S. M. van Anders. 2015. "Beyond Sexual Orientation: Integrating Gender/Sex and Diverse Sexualities in Sexual Configurations Theory." *Archives of Sexual Behavior* 44: 1177–213.

19. N. Krieger. 2019. "Measures of Racism, Sexism, Heterosexism, and Gender Binarism for Health Equity Research: From Structural Injustice to Embodied Harm—An Ecosocial Analysis." *Annual Review of Public Health* 41: 37–62; M. Yudell, D. Roberts, R. DeSalle, and S. Tishkoff. 2016. "Taking Race Out of Human Genetics." *Science* 351 (6273): 564–65. https://doi.org/10.1126/science.aac4951; M. Lock. 2015. "Comprehending the Body in the Era of the Epigenome." *Current Anthropology* 56 (2): 151–77; Fuentes, "Humans Are Biocultural"; Fuentes and Wiessner, "Reintegrating Anthropology"; Downey and Lende, "Evolution and the Brain"; Leatherman and Goodman, "Building on the Biocultural Syntheses"; Ingold and Paalson, *Biosocial Becomings*.

20. Here and in the following definitions I am drawing on a range of literature, but most specifically the excellent summaries of that literature and the definitions provided in Hyde et al., "Future of Sex and Gender"; L. Z. DuBois and H. Shattuck-Heidorn. 2021. "Challenging the Binary: Gender/Sex and the Bio-logics of Normalcy." *American Journal of Human Biology* 33 (5): e23623. https://doi.org/10.1002/ajhb.23623; van Anders, "Gender/Sex/ual Diversity and Biobehavioral Research"; and A. Fausto-Sterling. 2019. "Gender/Sex, Sexual Orientation, and Identity Are in the Body: How Did They Get There?" *The Journal of Sex Research* 56 (4–5): 529–55. https://doi.org/10.1080/00224499.2019.1581883; S. Richardson. 2013. *Sex Itself: The Search for Male and Female in the Human Genome*. Chicago: The University of Chicago Press.

21. D. Joel. 2012. "Genetic-Gonadal-Genitals Sex (3G-Sex) and the Misconception of Brain and Gender, or, Why 3G-Males and 3G-Females Have Intersex Brain and Intersex Gender." *Biology of Sex Differences* 3 (1): 27; DuBois and Shattuck-Heidorn, "Challenging the Binary."

22. The study might not explicitly state this, but it is almost always the case that when the researcher says "male" and "female," they mean 3G male and 3G female,

but in reality, researchers are almost always only taking the "sex at birth" assignment unless stated otherwise in the actual research work.

23. Most researchers assume they are using something like 3G categories, but in reality, they are often using self-reported "sex" (male/female) or "gender" (man/woman) or "sex" assessed based on physical observation of overall appearance. This means that there is usually some small error in the assignments and thus the data are often slightly askew.

24. M. Blackless, A. Charuvastra, A. Derryck, A. Fausto-Sterling, K. Lauzanne, and E. Lee. 2000. "How Sexually Dimorphic Are We? Review and Synthesis." *American Journal of Human Biology* 12 (2): 151–66.

4. Humans Then

1. Here I leave out the marsupials, who are very fascinating, but not as central to this discussion as are the placental mammals (as that is what humans are).

2. Relative to other nonmammal animals. Birds do have extended care, but not at the physiological and behavioral level of mammals.

3. S. Hrdy. 2009. *Mothers and Others: The Evolutionary Origins of Mutual Understanding*. Cambridge, MA: Harvard University Press; E. Fernandez-Duque, C. R. Valeggia, and S. P. Mendoza. 2009. "The Biology of Paternal Care in Human and Nonhuman Primates." *Annual Review of Anthropology* 38 (1): 115–30; S. Rosenbaum and J. B. Silk. 2022. "Pathways to Paternal Care in Primates." *Evolutionary Anthropology* 31 (5): 245–62. https://doi.org/10.1002/evan.21942.

4. Hrdy, *Mothers and Others*; C. Ross and A. MacLarnon. 2000. "The Evolution of Non-maternal Care in Anthropoid Primates: A Test of the Hypotheses." *Folia Primatol (Basel)* 71 (1–2): 93–113. https://doi.org/10.1159/000021733; S. Rosenbaum and L. T. Gettler. 2018. "With a Little Help from Her Friends (and Family) Part I: The Ecology and Evolution of Non-maternal Care in Mammals." *Physiology & Behavior* 193, Part A: 1–11. https://doi.org/10.1016/j.physbeh.2017.12.025.

5. E. Clarke, K. Bradshaw, K. Drissell, P. Kadam, N. Rutter, and S. Vaglio. 2022. "Primate Sex and Its Role in Pleasure, Dominance and Communication." *Animals* 12 (23): 3301. https://doi.org/10.3390/ani12233301.

6. A. Whiten, R. A. Hinde, K. N. Laland, and C. B. Stringer. 2011. "Culture Evolves." *Philosophical Transactions of the Royal Society of London: Biological Sciences* 366 (1567): 938–48. https://doi.org/10.1098/rstb.2010.0372; H. Whitehead, K. N. Laland, L. Rendell, R. Thorogood, and A. Whiten. 2019. "The Reach of Gene–Culture Coevolution in Animals." *Nature Communications* 10 (2405): https://doi.org/10.1038/s41467-019-10293-y.

7. J. M. Plavcan. 2012. "Sexual Size Dimorphism, Canine Dimorphism, and Male-Male Competition in Primates." *Human Nature* 23: 45–67; but also see A. Fuentes.

2021. "Searching for the 'Roots' of Masculinity in Primates and the Human Evolutionary Past." *Current Anthropology* 62: S13–S25. https://doi.org/10.1086/711582; and the discussion in this and the subsequent chapter in regard to humans in this context.

8. N. Smit, B. Ngoubangoye, M. J. E. Charpentier, and E. Huchard. 2022. "Dynamics of Intersexual Dominance in a Highly Dimorphic Primate." *Frontiers in Ecology and Evolution* 10: 931226. https://doi.org/10.3389/fevo.2022.931226.

9. P. M. Kappeler, E. Huchard, A. Baniel, C. Canteloup, M. J. E. Charpentier, L. Cheng, E. Davidian, et al. 2022. "Sex and Dominance: How to Assess and Interpret Intersexual Dominance Relationships in Mammalian Societies." *Frontiers in Ecology and Evolution* 10: 918773. https://doi.org/10.3389/fevo.2022.918773; R. J. Lewis. 2018. "Female Power in Primates and the Phenomenon of Female Dominance." *Annual Review of Anthropology* 47: 533–51. https://doi.org/10.1146/annurev-anthro-102317-045958; R. J. Lewis. 2020. "Female Power: A New Framework for Understanding 'Female Dominance' in Lemurs." *Folia Primatologica* 91: 48–68. https://doi.org/10.1159/000500443.

10. G. Parker. 2006. "Sexual Conflict over Mating and Fertilization: An Overview. *Philosophical Transactions of the Royal Society: Biological Sciences* 361: 235–59. https://doi.org/10.1098/rstb.20 05.1785.

11. C. Campbell, A. Fuentes, K. MacKinnon, S. Bearder, and R. Stumpf. 2011. *Primates in Perspective*. 2nd ed. Oxford: Oxford University Press, chap. 19.

12. Campbell et al., *Primates in Perspective*, chaps. 14 and 15.

13. L. T. Gettler. 2016. "Becoming DADS: Considering the Role of Cultural Context and Developmental Plasticity for Paternal Socioendocrinology." *Current Anthropology* 57: S38–S51; E. Fernandez-Duque, C. R. Valeggia, and S. P. Mendoza. 2009. "The Biology of Paternal Care in Human and Nonhuman Primates." *Annual Review of Anthropology* 38 (1): 115–30.

14. In fact, we sometimes see female dominance (such as in many lemurs).

15. T. Q. Bartlett and L. E. O. Light. "Sexual Dichromatism." In *The International Encyclopedia of Primatology*, edited by A. Fuentes. Hoboken: John Wiley & Sons, Inc. https://onlinelibrary.wiley.com/doi/10.1002/9781119179313.wbprim0427.

16. Campbell et al., *Primates in Perspective*, chap. 17.

17. A. Kralick. 2023. "When Ape Sex Isn't Simple." *Anthropology News*, March 14. https://www.anthropology-news.org/articles/when-ape-sex-isnt-simple/#citation.

18. A. N. Maggioncalda, R. M. Sapolsky, and N. M. Czekala. 1999. "Reproductive Hormone Profiles in Captive Male Orangutans: Implications for Understanding Developmental Arrest." *American Journal of Physical Anthropology* 109 (1): 19–32. https://doi.org/10.1002/(SICI)1096-8644(199905)109:1<19::AID-AJPA3>3.0.CO;2-3. And E. M. Thompson, A. Zhou, and C. D. Knott. 2012. "Low Testosterone Correlates with Delayed Development in Male Orangutans." *PloS One*. 7 (10): e47282. https://doi.org/10.1371/journal.pone.0047282.

19. Campbell et al., *Primates in Perspective*, chap. 18.

20. Campbell et al., *Primates in Perspective*, chap. 19.

21. S. Rosenbaum, L. Vigilant, C. W. Kuzawa, and T. S. Stoinski. 2018. "Caring for Infants Is Associated with Increased Reproductive Success for Male Mountain Gorillas." *Scientific Reports* 8: 15223.

22. Campbell et al., *Primates in Perspective*, chap. 20.

23. Campbell et al., *Primates in Perspective*, chap. 20.

24. A. Fuentes. 2018. "How Humans and Apes Are Different, and Why It Matters." *Journal of Anthropological Research* 74 (2): 151–67. https://doi.org/10.1086/697150; J. Marks. 2012. "The Biological Myth of Human Evolution." *Contemporary Social Science: Journal of the Academy of Social Sciences* 7 (2): 139–57.

25. Actually, none of the hominins have even the teeny canine dimorphism one sees in bonobos. G. Suwa, T. Sasaki, S. Semaw, M. J. Rogers, S. W. Simpson, Y. Kunimatsu, M. Nakatsukasa, et al. 2021. "Canine Sexual Dimorphism in *Ardipithecus ramidus* Was Nearly Human-Like." *Proceedings of the National Academy of Sciences of the United States of America* 118 (49): e2116630118. https://doi.org/10.1073/pnas.2116630118.

26. N. Malone, A. H. Wade, A. Fuentes, E. Riley, M. Remis, and C. Jost-Robinson. 2014. "Ethnoprimatology: Critical Interdisciplinarity and Multispecies Approaches in Anthropology." *Critique of Anthropology* 34 (1): 8–29.

27. S. Harmand, J. Lewis, and C. Feibel, C. J. Lepre, S. Prat, A. Lenoble, X. Boës, et al. 2015. "3.3-Million-Year-Old Stone Tools from Lomekwi 3, West Turkana, Kenya." *Nature* 521: 310–15. https://doi.org/10.1038/nature14464; A. Fuentes. 2017. *The Creative Spark: How Imagination Made Humans Exceptional*. New York: Dutton.

28. P. Frémondière, L. Thollon, F. Marchal, F. Marchal, C. Fornai, N. Webb, and M. Haeusler. 2022. "Dynamic Finite-Element Simulations Reveal Early Origin of Complex Human Birth Pattern." *Communications Biology* 5 (377): https://doi.org/10.1038/s42003-022-03321-z.

29. C. Kuzawa, H. T. Chugani, L. I. Grossman, L. Lipovich, O. Muzik, P. Hof, R. Patrick, D. E. Wildman, et al. 2014. "Metabolic Costs and Evolutionary Implications of Human Brain Development." *Proceedings of the National Academy of Sciences of the United States of America* 111 (36): 13010–15; K. R. Rosenberg. 2021. "The Evolution of Human Infancy: Why It Helps to Be Helpless." *Annual Review of Anthropology* 50: 423–40. https://doi.org/10.1146/annurev-anthro-111819-105454.

30. S. Lacy and C. Ocobock. 2023. "Woman the Hunter: The Archaeological Evidence." *American Anthropologist* 126 (1): 19–31. https://doi.org/10.1111/aman.13914.

31. R. A. Foley. 2016. "Mosaic Evolution and the Pattern of Transitions in the Hominin Lineage." *Philosophical Transactions of the Royal Society: Biological Sciences* 371: 1–14; Fuentes, *Creative Spark*; Fuentes, "How Humans and Apes Are Different";

J. Galway-Witham, J. Cole, and C. B. Stringer. 2019. "Aspects of Human Physical and Behavioural Evolution during the Last 1 Million Years." *Journal of Quaternary Science* 34 (6): 355–78.

32. Foley, "Mosaic Evolution"; Fuentes, *Creative Spark*; Fuentes, "How Humans and Apes Are Different"; Galway-Witham, Cole, and Stringer, "Aspects of Human Physical and Behavioural Evolution."

33. P. Spikins, A. Needham, B. Wright, C. Dytham, M. Gatta, and G. Hitchens. 2018. "Living to Fight Another Day: The Ecological and Evolutionary Significance of Neanderthal Healthcare." *Quaternary Science Reviews* 217: 98–118. https://doi.org/10.1016/j.quascirev.2018.08.011; P. Spikins, C. Dytham, J. French, and S. John-Wood. 2021. "Theoretical and Methodological Approaches to Ecological Changes, Social Behaviour and Human Intergroup Tolerance 300,000 to 30,000 BP." *Journal of Archaeological Method and Theory* 28: 53–75. https://doi.org/10.1007/s10816-020-09503-5.

34. Fuentes, *Creative Spark*; Fuentes, "How Humans and Apes Are Different"; S. Mithen. 2024. *The Language Puzzle: Piecing Together the Six Million Year Story of How Words Evolved*. New York: Basic Books.

35. Actually, allocare is pretty common in highly social mammals (like canids, some cats, mongooses, etc.).

36. Rosenbaum and Gettler, "With a Little Help from Her Friends Part I."

37. Rosenberg, "Evolution of Human Infancy."

38. Fuentes, *Creative Spark*; Fuentes, "How Humans and Apes Are Different"; Galway-Witham, Cole, and Stringer, "Aspects of Human Physical and Behavioural Evolution."

39. Hrdy, *Mothers and Others*.

40. L. T. Gettler. 2010. "Direct Male Care and Hominin Evolution: Why Male-Child Interaction Is More Than a Nice Social Idea." *American Anthropologist* 112: 7–21; A. M. Kubicka, R. Wragg Sykes, A. Nowell, and E. Nelson. 2022. "Sexual Behavior in Neanderthals." In *The Cambridge Handbook of Evolutionary Perspectives on Sexual Psychology*, 516–60. Cambridge: Cambridge University Press.

41. K. Hawkes and J. Coxworth. 2013. "Grandmothers and the Evolution of Human Longevity: A Review of Findings and Future Directions." *Evolutionary Anthropology* 22: 294–302. https://doi.org/10.1002/evan.21382.

42. Hrdy, *Mothers and Others*; Kuzawa et al., "Metabolic Costs"; Rosenberg, "Evolution of Human Infancy"; Gettler, "Direct Male Care and Hominin Evolution"; Gettler, "Becoming DADS."

43. P. Spikins. 2015. *How Compassion Made Us Human: The Evolutionary Origins of Tenderness, Trust & Morality*. Barnsley: Pen & Sword Archaeology; J. M. Burkart, S. B. Hrdy, and C. P. Van Schaik. 2009. "Cooperative Breeding and Human Cognitive Evolution." *Evolutionary Anthropology* 18 (5): 175–86; R. Sear. 2015. "Beyond the

Nuclear Family: An Evolutionary Perspective on Parenting." *Current Opinion in Psychology* 7: 98–103. https://doi.org.10.1016/j.copsyc.2015.08.013. Gettler, "Direct Male Care and Hominin Evolution"; Gettler, "Becoming DADS."

44. See case study overview for Neanderthals here: Kubicka et al., "Sexual Behavior in Neanderthals." See also P. K. Stone and A. Netzer Zimmer. 2023. "Issues of Gender, Identity, and Agency on Paleopathology." In *The Routledge Handbook of Paleopathology*, edited by A. L. Grauer, 435–56. Oxfordshire: Routledge.

45. G. R. Brown, K. N. Laland, and M. B. Mulder. 2009. "Bateman's Principles and Human Sex Roles." *Trends in Ecology and Evolution* 24 (6): 297–304. https://doi.org/ 10.1016/j.tree.2009.02.005.

46. Brown, Laland, and Mulder, "Bateman's Principles and Human Sex Roles," 297.

47. C. T. Ross, P. L. Hooper, J. E. Smith, A. V. Jaeggi, E. A. Smith, S. Gavrilets, F. T. Zohora, et al. 2023. "Reproductive Inequality in Humans and Other Mammals." *Proceedings of the National Academy of Sciences of the United States of America* 120 (22): e2220124120. https://doi.org/10.1073/pnas.2220124120.

48. H. Dunsworth. 2020. "Expanding the Evolutionary Explanations for Sex Differences in the Human Skeleton." *Evolutionary Anthropology: Issues, News, and Reviews* 29 (3): 108–16. https://doi.org/10.1002/evan.21834; A. Fuentes. 2016. "The Extended Evolutionary Synthesis, Ethnography, and the Human Niche: Toward an Integrated Anthropology." *Current Anthropology* 57 (suppl. 13): 13–26. https://doi.org/10.1086/685684; Gettler, "Direct Male Care and Hominin Evolution"; Sear, "Beyond the Nuclear Family"; R. Sear. 2021. "The Male Breadwinner Nuclear Family Is Not the 'Traditional' Human Family, and Promotion of This Myth May Have Adverse Health Consequences." *Philosophical Transactions of the Royal Society: Biological Sciences* 376 (1827): 20200020. https://doi.org/10.1098/rstb.2020.0020.

49. Clarke et al., "Primate Sex and Its Role in Pleasure."

50. M. Rodrigues. 2021. "Same Sex Sexual Behavior in Chimpanzees Challenge Our Gendered Biases about Evolution." *Prosocial World*, accessed September 8, 2024. https://thisviewoflife.com/same-sex-sexual-behavior-in-chimpanzees-challenge-our-gendered-biases-about-evolution/.

51. There are four subspecies of chimpanzees, and all have been reported to engage in sexual behavior outside of reproduction, but not at the levels seen in bonobos.

52. R. Stumpf. 2010. "Chimpanzees and Bonobos: Inter- and Intra-species Diversity." In *Primates in Perspective*, 2nd ed., edited by C. Campbell, A. Fuentes, K. MacKinnon, S. Bearder, and R. Stumpf, 340–56. Oxford: Oxford University Press; A. Sandel and R. Reddy. 2021. "Sociosexual Behaviour in Wild Chimpanzees Occurs in Variable Contexts and Is Frequent between Same-Sex Partners." *Behaviour* 158: 1–28. https://doi.org.10.1163/1568539X-bja10062.

53. G. Hohmann and B. Fruth. 2000. "Use and Function of Genital Contacts among Female Bonobos." *Animal Behavior* 60: 107–20. https://doi.org.10.1006/anbe.2000.1451; L. R. Moscovice, M. Surbeck, B. Fruth, G. Hohmann, A. V. Jaeggi, and T. Deschner. 2019. "The Cooperative Sex: Sexual Interactions among Female Bonobos Are Linked to Increases in Oxytocin, Proximity and Coalitions." *Hormones and Behavior* 116: 104581. https://doi.org/10.1016/j.yhbeh.2019.104581.

54. Sandel and Reddy, "Sociosexual Behaviour in Wild Chimpanzees."

55. Sandel and Reddy, "Sociosexual Behaviour in Wild Chimpanzees"; J. Yamagiwa. 2006. "Playful Encounters: The Development of Homosexual Behaviour in Male Mountain Gorillas." In *Homosexual Behaviour in Animals: An Evolutionary Perspective*, edited by V. Sommer and P. L. Vasey, 273–93. Cambridge: Cambridge University Press.

56. E. A. Fox. 2001. "Homosexual Behavior in Wild Sumatran Orangutans (*Pongo pygmaeus abelii*)." *American Journal of Primatology* 55: 177–81. https://doi.org/10.1002/ajp.1051.

57. A. Fuentes. 2022. *Race, Monogamy and Other Lies They Told You: Busting Myths about Human Nature*. 2nd ed. Oakland: University of California Press; K. Wellings, M. Collumbien, E. Slaymaker, S. Singh, Z. Hodges, D. Patel, and N. Bajos. 2006. "Sexual Behaviour in Context: A Global Perspective." *Lancet* 368 (9548): 1706–28. https://doi.org/10.1016/S0140-6736(06)69479-8.

58. Again, for a case study in Neanderthals, see Kubicka et al., "Sexual Behavior in Neanderthals."

59. O. Lovejoy. [1981] 2009. "Reexamining Human Origins in Light of *Ardipithecus ramidus*." *Science* 326: 108–15.

60. Fuentes, *Race, Monogamy and Other Lies*.

61. For recent versions of this approach, see B. Chapais. 2010. *Primeval Kinship: How Pair-Bonding Gave Birth to Human Society*. Cambridge, MA: Harvard University Press; and Lovejoy, "Reexamining Human Origins"; see also Ross et al., "Reproductive Inequality in Humans and Other Mammals."

62. B. Chapais. 2010. *Primeval Kinship: How Pair-Bonding Gave Birth to Human Society*. Cambridge, MA: Harvard University Press.

63. Fuentes, *Race, Monogamy and Other Lies*.

64. A. Fuentes. 2009. *Evolution of Human Behavior*. Oxford: Oxford University Press; and A. Fuentes, M. Wyczalkowski, and K. C. MacKinnon. 2010. "Niche Construction through Cooperation: A Nonlinear Dynamics Contribution to Modeling Facets of the Evolutionary History in the Genus *Homo*." *Current Anthropology* 51 (3): 435–44; Fuentes *Creative Spark*.

65. See A. Fuentes. 1999. "Re-evaluating Primate Monogamy." *American Anthropologist* 100 (4): 890–907; and A. Fuentes. 2002. "Patterns and Trends in Primate Pair Bonds." *International Journal of Primatology* 23 (4): 953–78.

66. Less than 3 percent of all mammalian species are monogamous.

67. See multiple chapters in P. T. Ellison and P. B. Gray, eds. 2009. *The Endocrinology of Social Relationships*, 270–93. Cambridge, MA: Harvard University Press; and J. T. Curtis and Z. Wang. 2003. "The Neurochemistry of Pair Bonding." *Current Directions in Psychological Science* 12 (2): 49–53; and S. L. Díaz-Muñoz, and K. L. Bales. 2016. "Monogamy." In "Primates: Variability, Trends, and Synthesis." Introduction to special issue on Primate Monogamy. *American Journal of Primatology* 78 (3): 283–87. https://doi.org/10.1002/ajp.22463.

68. Fuentes, *Race, Monogamy and Other Lies.*

69. See multiple chapters in Ellison and Gray, *Endocrinology of Social Relationships*, 270–93; and Curtis and Wang, "Neurochemistry of Pair Bonding."

70. However, just being a relative does not automatically generate a pair bond.

71. Fuentes, *Creative Spark*; M. W. Conkey. 2010. "Images without Words: The Construction of Prehistoric Imaginaries for Definitions of 'Us.'" *Journal of Visual Culture* 9: 272–83; M. Kissel and A. Fuentes. 2017. "Semiosis in the Pleistocene." *Cambridge Archaeological Journal* 27 (3): 1–16. https://doi.org/10.1017/S0959774317000014.

72. S. Nanda. 2014. *Gender Diversity: Cross-Cultural Variations.* 2nd ed. Long Grove, IL: Waveland Press; W. Wood and A. H. Eagly. 2002. "A Cross-Cultural Analysis of the Behavior of Women and Men: Implications for the Origins of Sex Differences." *Psychological Bulletin* 128 (5): 699–727; J. Lorber. 2022. *The New Gender Paradox.* Oxford: Polity Press.

73. S. Lacy and C. Ocobock. 2023. "Woman the Hunter: The Archaeological Evidence." *American Anthropologist* 126 (1): 19–31. https://doi.org/10.1111/aman.13914.

74. L. C. K. Aiello. 2002. "Energetic Consequences of Being a *Homo erectus* Female." *American Journal of Human Biology* 14: 551–65; L. C. Aiello and J. C. K. Wells. 2002. "Energetics and the Evolution of the Genus *Homo*." *Annual Review of Anthropology* 31: 323–38.

75. D. Haraway. 1989. *Primate Visions: Gender, Race, and Nature in the World of Modern Science.* Oxfordshire: Routledge; Fuentes, *Creative Spark*; Conkey, "Images without Words."

76. Lacy and Ocobock, "Woman the Hunter: The Archaeological Evidence."

77. Stone and Zimmer, "Issues of Gender, Identity, and Agency on Paleopathology"; Fuentes, *Creative Spark*; Lacy and Ocobock, "Woman the Hunter: The Archaeological Evidence"; C. Ocobock and S. Lacy. 2023. "Woman the Hunter: The Physiological Evidence." *American Anthropologist* 126 (1): 7–18. https://doi.org/10.1111/aman.13915.

78. See J. M. Adovasio, O. Soffer, and J. Page. 2007. *The Invisible Sex: Uncovering the True Roles of Women in Prehistory.* Washington, DC: Smithsonian Books; Stone and Zimmer, "Issues of Gender, Identity, and Agency on Paleopathology"; Conkey, "Images without Words."

79. D. Stout. 2011. "Stone Toolmaking and the Evolution of Human Culture and Cognition." *Philosophical Transactions of the Royal Society B* 366 (1567): 1050–59; D. Stout and N. Khreisheh. 2015. "Skill Learning and Human Brain Evolution: An Experimental Approach." *Cambridge Archaeological Journal* 25: 867–75. https://doi.org/10.1017/S0959774315000359; Lacy and Ocobock, "Woman the Hunter: The Archaeological Evidence."

80. K. W. Arthur. 2010. "Feminine Knowledge and Skill Reconsidered: Women and Flaked Stone Tools." *American Anthropologist* 112 (2): 228–43; J. Gero. 1991. "Genderlithics: Women's Roles in Stone-Tool Production." In *Engendering Archaeology: Women and Prehistory*, edited by J. Gero and M. W. Conkey, 163–93. Oxford: Blackwell; K. Weedman. 2005. "Gender and Stone Tools: An Ethnographic Study of the Konso and Gamo Hideworkers of Southern Ethiopia." In *Gender and Hide Production*, edited by L. Frink and K. Weedman, 175–97. Walnut Creek, CA: Altamira Press.

81. A. Murray, R. Pinhasi, and J. Stock. 2014. "Lower Limb Skeletal Biomechanics Track Long-Term Decline in Mobility across ~6150 Years of Agriculture in Central Europe." *Journal of Archaeological Science* 52: 376–90. https://doi.org/10.1016/j.jas.2014.09.001. A. A. Macintosh, R. Pinhasi, and J. T. Stock. 2017. "Prehistoric Women's Manual Labor Exceeded That of Athletes through the First 5500 Years of Farming in Central Europe." *Science Advances* 3 (11): eaao3893. https://doi.org/10.1126/sciadv.aao3893; Lacy and Ocobock, "Woman the Hunter: The Archaeological Evidence"; Ocobock and Lacy, "Woman the Hunter: The Physiological Evidence."

82. Ocobock and Lacy, "Woman the Hunter: The Physiological Evidence."

83. E.g. Lacy and Ocobock, "Woman the Hunter: The Archaeological Evidence"; Ocobock and Lacy, "Woman the Hunter: The Physiological Evidence"; R. Haas, J. Watson, T. Buonasera, J. Southon, J. C. Chen, S. Noe, K. Smith, et al. 2020. "Female Hunters of the Early Americas." *Science Advances* 6 (45): eabd0310. https://doi.org/10.1126/sciadv.abd0310; S. L. Kuhn and M. C. Stiner. 2006. "What's a Mother to Do? The Division of Labor among Neandertals and Modern Humans in Eurasia." *Current Anthropology* 47 (6): 953–80. https://doi.org/10.1086/507197; R. Wragg Sykes. 2020. *Kindred: Neanderthal Life, Love, Death and Art*. London: Bloomsbury; A. A. Sigma, S. Chilczuk, K. Nelson, R. Ruther, and C. Wall-Scheffler. 2023. "The Myth of Man the Hunter: Women's Contribution to the Hunt across Ethnographic Contexts." *PloS One* 18 (6): e0287101. https://doi.org/10.1371/journal.pone.0287101.

84. Lacy and Ocobock, "Woman the Hunter: The Archaeological Evidence"; Ocobock and Lacy, "Woman the Hunter: The Physiological Evidence."

85. A. Estalrrich and A. Rosas. 2015. "Division of Labor by Sex and Age in Neandertals: An Approach through the Study of Activity-Related Dental Wear." *Journal*

of Human Evolution 80: 51–63; see also Kubicka et al., "Sexual Behavior in Neanderthals."

86. Spikins et al., "Living to Fight Another Day"; Spikins et al., "Theoretical and Methodological Approaches."

87. D. Snow. 2013. "Sexual Dimorphism in European Upper Paleolithic Cave Art." *American Antiquity* 78: 746–61. https://doi.org/10.7183/0002-7316.78.4.746; Conkey, "Images without Words"; Lacy and Ocobock, "Woman the Hunter: The Archaeological Evidence."

88. P. Sorokowski and M. Kowal. 2024. "Relationship between the 2D:4D and Prenatal Testosterone, Adult Level Testosterone, and Testosterone Change: Meta-analysis of 54 Studies." *American Journal of Biological Anthropology* 183 (1): 20–38. https://doi.org/10.1002/ajpa.24852; M. Marczak, M. Misiak, A. Sorokowska, and P. Sorokowski. 2018. "No Sex Difference in Digit Ratios (2D:4D) in the Traditional Yali of Papua and Its Meaning for the Previous Hypotheses on the Inter-populational Variability in 2D:4D." *American Journal of Human Biology* 30 (2): https://doi.org/10.1002/ajhb.23078.

89. B. McCauley, D. Maxwell, and M. Collard. 2018. "A Cross-cultural Perspective on Upper Palaeolithic Hand Images with Missing Phalanges." *Journal of Paleolithic Archeology* 1: 314–33. https://doi.org/10.1007/s41982-018-0016-8.

90. Fuentes, *Creative Spark*; A. Fuentes. 2019. *Why We Believe: Evolution and the Human Way of Being*. New Haven, CT: Yale University Press/Templeton Press; D. Graeber and D. Wengrow. 2021. *The Dawn of Everything: A New History of Humanity*. New York: Farrar, Strauss and Giroux.

91. Fuentes, *Creative Spark*; Lacy and Ocobock, "Woman the Hunter: The Archaeological Evidence"; Ocobock and Lacy, "Woman the Hunter: The Physiological Evidence."

92. For example, M. Cintas-Peña and L. García Sanjuán. 2019. "Gender Inequalities in Neolithic Iberia: A Multi-Proxy Approach." *European Journal of Archaeology* 22 (4): 499–522. https://doi.org/10.1017/eaa.2019.3; R. Rasteiro and L. Chikhi. 2013. "Female and Male Perspectives on the Neolithic Transition in Europe: Clues from Ancient and Modern Genetic Data." *PloS One* 8 (4): e60944. https://doi.org/10.1371/journal.pone.0060944.

93. C. Hedenstierna-Jonson, A. Kjellström, T. Zachrisson, M. Krzenwińska, V. Sobrado, N. Price, T. Günther, M. Jakobsson, et al. 2017. "A Female Viking Warrior Confirmed by Genomics." *American Journal of Physical Anthropology* 164: 853–60. https://doi.org/10.1002/ajpa.23308; K. Croucher. 2012. *Death and Dying in the Neolithic Near East*. Oxford: Oxford University Press; M. Cintas-Peña, M. Luciañez-Triviño, R. Montero Artús, A. Bileck, P. Bortel, F. Kanz, K. Rebay-Salisbury, et al. 2023. "Amelogenin Peptide Analyses Reveal Female Leadership in Copper Age Iberia (c. 2900–2650 BC)." *Scientific Reports* 13 (1): 9594. https://doi.org/10.1038/s41598-023-36368-x.

5. Humans Now

1. Note I am specifically stating male and female (3G), not man and woman . . . if one used only gender categories, there would be even more mixing and less consistency between one of the categories and height.

2. A. Fausto-Sterling. 2020. *Sexing the Body: Gender Politics and the Construction of Sexuality*. London: Hachette.

3. See for example C. Sanchis-Segura and R. R. Wilcox. 2024. "From Means to Meaning in the Study of Sex/Gender Differences and Similarities." *Front Neuroendocrinol* 73: 101133. https://doi.org/10.1016/j.yfrne.2024.101133.

4. C. Ruff. 2002. "Variation in Human Body Size and Shape." *Annual Reviews in Anthropology* 31: 211–32; A. Fuentes. 2022. *Race, Monogamy and Other Lies They Told You: Busting Myths about Human Nature*. 2nd ed. Oakland: University of California Press.

5. Fausto-Sterling, *Sexing the Body*; L. Z. DuBois and H. Shattuck-Heidorn. 2021. "Challenging the Binary: Gender/Sex and the Bio-logics of Normalcy." *American Journal of Human Biology* 33 (5): e23623; Fuentes, *Race, Monogamy and Other Lies*; W. D. Lassek and S. J. C. Gaulin. 2022. "Substantial but Misunderstood Human Sexual Dimorphism Results Mainly from Sexual Selection on Males and Natural Selection on Females." *Frontiers in Psychology* 17 (13): 859931. https://doi.org/10.3389/fpsyg.2022.859931.

6. See the National Center for Health Statistics. https://www.cdc.gov/nchs/nhanes/index.htm.

7. C. D. Fryar, M. D. Carroll, Q. Gu, J. Afful, and C. L. Ogden. 2021. "Anthropometric Reference Data for Children and Adults: United States, 2015–2018." *Vital Health Statistics* (36): 1–44.

8. H. M. Dunsworth. 2020. "Expanding the Evolutionary Explanations for Sex Differences in the Human Skeleton." *Evolutionary Anthropology* 29 (3): 108–16. https://doi.org/10.1002/evan.21834.

9. T. M. Flaherty, L. J. Johnson, K. C. Woollen, D. Lopez, K. Gaddis, S. L. Horsley, and J. F. Byrnes. 2023. "Speaking of Sex: Critical Reflections for Forensic Anthropologists." *Humans* 3: 251–70. https://doi.org/10.3390/humans3040020; E. Garofalo and H. Garvin. 2020. "The Confusion between Biological Sex and Gender and Potential Implications of Misinterpretations." https://doi.org/10.1016/B978-0-12-815767-1.00004-3; R. M. Meloro, S. D. Tallman, C. G. Streed Jr., J. Stowell, T. A. Delgado, J. D. Haug, A. Redgrave, et al. Forthcoming. "A Framework for Incorporating Diverse Gender Identities into Forensic Anthropology Casework and Theory: Recommendations for Inclusive Practices." *Current Anthropology*.

10. J. L. Nuzzo. 2022. "Narrative Review of Sex Differences in Muscle Strength, Endurance, Activation, Size, Fiber Type, and Strength Training Participation Rates,

Preferences, Motivations, Injuries, and Neuromuscular Adaptations." *Journal of Strength and Conditioning Research* 37 (2): 494–536. https://doi.org/10.1519/JSC.0000000000004329.

11. Interestingly, the difference is in concentric rather than eccentric muscle contractions.

12. A. E. J. Miller, J. D. MacDougall, M. A. Tarnopolsky, and D. G. Sale. 1993. "Gender Differences in Strength and Muscle Fiber Characteristics." *European Journal of Applied Physiology and Occupational Physiology* 66 (3): 254–62. https://doi.org/10.1007/BF00235103; and see overview in C. Ocobock and S. Lacy. 2023. "Woman the Hunter: The Physiological Evidence." *American Anthropologist* 126 (1): 7–18. https://doi.org/10.1111/aman.13915.

13. P. Bishop, K. Cureton, and M. Collins. 1987. "Sex Difference in Muscular Strength in Equally-Trained Men and Women." *Ergonomics* 30 (4): 675–87. https://doi.org/10.1080/00140138708969760.

14. R. Kataoka, R. W. Spitz, V. Wong, Z. W. Bell, Y. Yamada, J. S. Song, W. B. Hammert, et al. 2023. "Sex Segregation in Strength Sports: Do Equal-Sized Muscles Express the Same Levels of Strength between Sexes?" *American Journal of Human Biology* 35 (5): e23862. https://doi.org/10.1002/ajhb.23862.

15. S. Welle, R. Tawil, and C. A. Thornton. 2008. "Sex-Related Differences in Gene Expression in Human Skeletal Muscle." *PloS One* 3 (1): e1385. https://doi.org/10.1371/journal.pone.0001385; Kataoka et al., "Sex Segregation in Strength Sports."

16. See overview in Ocobock and Lacy, "Woman the Hunter: The Physiological Evidence"; R. Jordan-Young and K. Karkazis. 2019. *Testosterone: An Unauthorized Biography*. Cambridge, MA: Harvard University Press.

17. N. Davis. 2023. "Speed, Angle and Confidence: Science behind Chloe Kelly's Powerful Penalty." *The Guardian*, August 11. https://www.theguardian.com/football/2023/aug/11/science-chloe-kelly-penalty-england-nigeria.

18. L. Zaccagni and E. Gualdi-Russo. 2023. "The Impact of Sports Involvement on Body Image Perception and Ideals: A Systematic Review and Meta-Analysis." *International Journal of Environmental Research and Public Health* 20 (6): 5228. https://doi.org/10.3390/ijerph20065228.

19. J. Henrich, S. J. Heine, and A. Norenzayan. 2010. "The Weirdest People in the World?" *Behavioral and Brain Sciences* 33 (2–3): 61–83, discussion 83–135. https://doi.org/10.1017/S0140525X0999152X.

20. A. Dreger. 2000. *Hermaphrodites and the Medical Invention of Sex*. Cambridge, MA: Harvard University Press; V. Sanz. 2017. "No Way Out of the Binary: A Critical History of the Scientific Production of Sex." *Signs: Journal of Women in Culture and Society* 43 (1): 1–27. https://doi.org/10.1086/692517.

21. Fausto-Sterling, *Sexing the Body*; R. O. Prum. 2023. *Performance All the Way Down: Genes, Development and Sexual Difference*. Chicago: The University of Chicago Press.

22. L. Baskin, J. Shen, A. Sinclair, M. Cao, X. Liu, G. Liu, D. Isaacson, et al. 2018. "Development of the Human Penis and Clitoris." *Differentiation* 103: 74–85. https://doi.org/10.1016/j.diff.2018.08.001; Prum, *Performance All the Way Down*; L. Boulanger, M. Pannetier, L. Gall, A. Allais-Bonnet, M. Elzaiat, D. Le Bourhis, N. Daniel, et al. 2014. "E. FOXL2 Is a Female Sex-Determining Gene in the Goat." *Current Biology* 24 (4): 404–8. https://doi.org/10.1016/j.cub.2013.12.039; F. Zhao, H. L. Franco, K. F. Rodriguez, P. R. Brown, M. Tsai, S. Y. Tsai, and H. H-C. Yao. 2017. "Elimination of the Male Reproductive Tract in the Female Embryo Is Promoted by COUP-TFII in Mice." *Science* 357: 717–20. https://doi.org/10.1126/science.aai9136.

23. Fausto-Sterling, *Sexing the Body*.

24. Specifically, the lack of anti-Müllerian hormone (AMH) and the correct sequence of activation of *EMX2, HOXA13, PAX2, LIM1*, and *WNT* genes. D. Wilson and B. Bordoni. 2023. "Embryology, Mullerian Ducts (Paramesonephric Ducts)." StatPearls [Internet], March 6. https://www.ncbi.nlm.nih.gov/books/NBK557727/.

25. T. Kurita. 2010. "Developmental Origin of Vaginal Epithelium." *Differentiation* 80 (2–3): 99–105. https://doi.org/10.1016/j.diff.2010.06.007; C. R. Cunha, S. J. Robboy, T. Kurita, D. Isaacson, J. Shen, M. Cao, and L. S. Baskin. 2018. "Development of the Human Female Reproductive Tract." *Differentiation* 103: 46–65. https://doi.org/10.1016/j.diff.2018.09.001.

26. M. Habiba, R. Heyn, P. Bianchi, I. Brosens, and G. Benagiano. 2021. "The Development of the Human Uterus: Morphogenesis to Menarche." *Human Reproduction Update* 27 (1): 1–26. https://doi.org/10.1093/humupd/dmaa036.

27. Again, see Fausto-Sterling, *Sexing the Body*, for an excellent overview of these patterns and the medical, and cultural, responses to them. See also many sites such as the Cleveland Clinic for public-facing information on this: "Atypical Genitalia (Formerly Known as Ambiguous Genitalia)." March 9, 2022. https://my.clevelandclinic.org/health/diseases/22470-atypical-genitalia-formerly-known-as-ambiguous-genitalia.

28. L. Scheja and J. Heeren. 2019. "The Endocrine Function of Adipose Tissues in Health and Cardiometabolic Disease." *Nature Reviews Endocrinology* 15 (9): 507–24. https://doi.org/10.1038/s41574-019-0230-6; C. SturtzSreetharan, A. Brewis, J. Hardin, S. Trainer, and A. Wutich. 2021. *Fat in Four Cultures: A Global Ethnography of Weight*. Toronto: University of Toronto Press.

29. T. Tchkonia, T. Thomou, Y. Zhu, I. Karagiannides, C. Pothoulakis, M. D. Jensen MD, and J. L. Kirkland. 2013. "Mechanisms and Metabolic Implications of Regional Differences among Fat Depots." *Cell Metabolism* 17 (5): 644–56. https://doi.org/10.1016/j.cmet.2013.03.008; N. Boulet, A. Briot, J. Galitzky, and A. Bouloumié. 2022. "The Sexual Dimorphism of Human Adipose Depots." *Biomedicines* 10 (10): 2615. https://doi.org/10.3390/biomedicines10102615.

30. R.W. Taylor, A. M. Grant, S. M. Williams, and A. Goulding. 2010. "Sex Differences in Regional Body Fat Distribution from Pre-to Postpuberty." *Obesity* 18: 1410–16.

31. C. M. Pond. 1997. "The Biological Origins of Adipose Tissue in Humans." In *The Evolving Female*, edited by M. E. Morbeck, A. Galloway, and A. L. Zihlman, 147–62. Princeton, NJ: Princeton University Press; Boulet et al., "Sexual Dimorphism of Human Adipose Depots."

32. A. Innocenti, D. Melita, and E. Dreassi. 2022. "Incidence of Complications for Different Approaches in Gynecomastia Correction: A Systematic Review of the Literature." *Aesthetic Plastic Surgery* 46: 1025–41. https://doi.org/10.1007/s00266-022-02782-1; D. Joel. 2012. "Genetic-Gonadal-Genitals Sex (3G-Sex) and the Misconception of Brain and Gender, or, Why 3G-Males and 3G-Females Have Intersex Brain and Intersex Gender." *Biology of Sex Differences* 3 (1): 27.

33. Boulet et al., "Sexual Dimorphism of Human Adipose Depots."

34. The variation in hair patterns across our species has been mispresented and misused by racists and specifically "race scientists" to make assertions about the lack or preponderance of hair on bodies and their relative "value." These same biases have been used to discriminate against bodies with "more" hair than culturally "appropriate." E. Carlin and B. Kramer. 2020. "Hair, Hormones, and Haunting: Race as a Ghost Variable in Polycystic Ovary Syndrome." *Science, Technology, & Human Values* 45 (5): 779–803. https://doi.org/10.1177/0162243920908647.

35. See B. O. Yildiz, S. Bolour, K. Woods, A. Moore, and R. Azziz. 2010. "Visually Scoring Hirsutism." *Human Reproduction Update* 16 (1): 51–64. https://doi.org/10.1093/humupd/dmp024. For some methods and data, but be aware of the racializing frame those authors see. See also E. Carlin and Kramer, "Hair, Hormones, and Haunting."

36. See for example T. Lasisi, J. W. Smallcombe, W. L. Kenney, M. D. Shriver, B. Zydney, N. G. Jablonski, and G. Havenith. 2023. "Human Scalp Hair as a Thermoregulatory Adaptation." *Proceedings of the National Academy of Sciences of the United States of America* 120 (24): e2301760120. https://doi.org/10.1073/pnas.2301760120; and also U. Ohnemus, M. Uenalan, J. Inzunza, J. A. Gustafsson, and R. Paus. 2006. "The Hair Follicle as an Estrogen Target and Source." *Endocrine Reviews* 27 (6): 677–706. https://doi.org/10.1210/er.2006-0020.

37. A. Melk, B. Babitsch, B. Borchert-Mörlins, F. Claas, A. I. Dipchand, S. Eifert, B. Eiz-Vesper, et al. 2019. "Equally Interchangeable? How Sex and Gender Affect Transplantation." *Transplantation* 103 (6): 1094–110. https://doi.org/10.1097/TP.0000000000002655; F. Puoti, A. Ricci, A. Nanni-Costa, W. Ricciardi, W. Malorni, and E. Ortona. 2016. "Organ Transplantation and Gender Differences: A Paradigmatic Example of Intertwining between Biological and Sociocultural Determinants."

Biological Sex Differerences 7: 35. https://doi.org/10.1186/s13293-016-0088-4; C. N. Bairey Merz, L. M. Dember, J. R. Ingelfinger, A. Vinson, J. Neugarten, K. L. Sandberg, J. C. Sullivan, et al. 2019. "Sex and the Kidneys: Current Understanding and Research Opportunities." *Nature Reviews Nephrology* 15 (12): 776–83. https://doi.org/10.1038/s41581-019-0208-6.

38. Prum, *Performance All the Way Down.*

39. P. T. Ellison. 2001. *On Fertile Ground.* Cambridge, MA: Harvard University Press.

40. Jordan-Young and Karkazis, *Testosterone*; J. S. Williams, M. R. Fattori, I. R. Honeyborne, and S. A. Ritz. 2023. "Considering Hormones as Sex- and Gender-Related Factors in Biomedical Research: Challenging False Dichotomies and Embracing Complexity." *Hormones and Behavior* 156: 105442. https://doi.org/10.1016/j.yhbeh.2023.105442.

41. As do most mammals and many other animals.

42. P. T. Ellison and P.B. Gray. 2009. *The Endocrinology of Social Relationships.* Cambridge, MA: Harvard University Press; R. G. Bribiescas. 2001. "Reproductive Ecology and Life History of the Human Male." *Yearbook of Physical Anthropology* 44: 148–76; Jordan-Young and Karkazis, *Testosterone*; S. M. van Anders, J. Steiger, and K. L. Goldey. 2015. "Effects of Gendered Behavior on Testosterone in Women and Men." *Proceedings of the National Academy of Sciences of the United States of America* 112: 13805–10; J. S. Williams, M. R. Fattori, I. R. Honeyborne, and S. A. Ritz. 2023. "Considering Hormones as Sex- and Gender-Related Factors in Biomedical Research: Challenging False Dichotomies and Embracing Complexity." *Hormones and Behavior* 156: 105442. https://doi.org/10.1016/j.yhbeh.2023.105442.

43. These hormones are also produced at lower levels by the adrenals and areas of the brain, and can be generated though chemical reactions in adipose tissue and at a few other sites in the body. But the gonads are the main producers of these three specific hormone types. The gonads also produce a few other hormones such as anti-Mullerian hormone (in ovaries) and inhibin (in all gonads).

44. Remember, there are many animals that have one gonad type: an ovotestis.

45. M. R. Bell. 2018. "Comparing Postnatal Development of Gonadal Hormones and Associated Social Behaviors in Rats, Mice, and Humans." *Endocrinology* 159 (7): 2596–613. https://doi.org/10.1210/en.2018-00220.

46. Fausto-Sterling, *Sexing the Body*; J. S. Hyde, R. S. Bigler, D. Joel, C. C. Tate, and S. M. van Anders. 2018. "The Future of Sex and Gender in Psychology: Five Challenges to the Gender Binary." *American Psychologist* 74 (2): 171–93. http://dx.doi.org/10.1037/amp0000307.

47. L. Lanciotti, M. Cofini, A. Leonardi, L. Penta, and S. Esposito. 2018. "Up-To-Date Review About Minipuberty and Overview on Hypothalamic-Pituitary-Gonadal

Axis Activation in Fetal and Neonatal Life." *Front Endocrinol (Lausanne)* 9: 410. https://doi.org/10.3389/fendo.2018.00410.

48. R. L. Rosenfield. 2021. "Normal and Premature Adrenarche." *Endocrine Reviews* 42 (6): 783–814. https://doi.org/10.1210/endrev/bnab009.

49. S. Feldman Witchel and A. Kemal Topaloglu. 2019. "Puberty: Gonadarche and Adrenarche." In *Yen and Jaffe's Reproductive Endocrinology*, 8th ed., edited by J. F. Strauss and R. L. Barbieri, 394–445. Amsterdam: Elsevier.

50. K. Peacock and K. M. Ketvertis 2022. *Menopause*. StatPearls Publishing. https://www.ncbi.nlm.nih.gov/books/NBK507826/; K. Clancy. 2023. *Period: The Real Story of Menstruation*. Princeton, NJ: Princeton University Press.

51. A. M. Matsumoto. 2002. "Andropause: Clinical Implications of the Decline in Serum Testosterone Levels with Aging in Men." *Journal of Gerontology: Biological Sciences* 57 (2): M76–99. https://doi.org/10.1093/gerona/57.2.m76.

52. M. Liu, S. Murthi, and L. Poretsky. 2020. "Polycystic Ovary Syndrome and Gender Identity." *Yale Journal of Biology and Medicine* 93 (4): 529–37.

53. C. Hooven. 2021. *T: The Story of Testosterone, the Hormone That Dominates and Divides Us*. New York: Henry Holt and Company.

54. Van Anders, Steiger, and Goldey, "Effects of Gendered Behavior"; K. O. Smiley, K. M. Munley, K. Aghi, S. E. Lipshutz, T. M. Patton, D. S. Pradhan, T. K. Solomon-Lane, et al. 2024. "Sex Diversity in the 21st Century: Concepts, Frameworks, and Approaches for the Future of Neuroendocrinology." *Hormones and Behavior* 157: 105445. https://doi.org/10.1016/j.yhbeh.2023.105445; K. Casto, D. Cohen, M. Akinola, and P. Mehta. 2024. "Testosterone, Gender Identity and Gender-Stereotyped Personality Attributes." *Hormones and Behavior* 162: 105540. https://doi.org/10.1016/j.yhbeh.2024.105540.

55. Jordan-Young and Karkazis, *Testosterone*; C. Fine. 2016. *Testosterone Rex: Myths of Sex, Science and Society*. New York: W. W. Norton; van Anders, Steiger, and Goldey, "Effects of Gendered Behavior"; Smiley et al., "Sex Diversity in the 21st Century"; Casto et al., "Testosterone, Gender Identity and Gender-Stereotyped Personality Attributes."

56. M. Schulster, A. M. Bernie, and R. Ramasamy. 2016. "The Role of Estradiol in Male Reproductive Function." 2016. *Asian Journal of Andrology* 18 (3): 435–40. https://doi.org/10.4103/1008-682X.173932; S. M. van Anders. 2013. "Beyond Masculinity: Testosterone, Gender/Sex, and Human Social Behavior in a Comparative Context." *Frontiers in Neuroendocrinology* 34: 198–210. https://doi.org/10.1016/j.yfrne.2013.07.001; Jordan-Young and Karkazis, *Testosterone*.

57. J. R. Roney. 2016. "Theoretical Frameworks for Human Behavioral Endocrinology." *Hormones and Behavior* 84: 97–110; Bribiescas, "Reproductive Ecology"; L. T. Gettler. 2014. "Applying Socioendocrinology to Evolutionary Models:

Fatherhood and Physiology." *Evolutionary Anthropology* 23: 146–60; L. T. Gettler. 2016. "Becoming DADS: Considering the Role of Cultural Context and Developmental Plasticity for Paternal Socioendocrinology." *Current Anthropology* 57: S38–S51; van Anders, Steiger, and Goldey, "Effects of Gendered Behavior"; S. M. van Anders, K. L. Goldey, and S. N. Bell. 2014. "Measurement of Testosterone in Human Sexuality Research: Methodological Donsiderations." *Archives of Sexual Behavior* 43: 231–50. http://doi.org/10.1007/s10508-013-0123-z.

58. van Anders, "Beyond Masculinity"; van Anders, Steiger, and Goldey, "Effects of Gendered Behavior."

59. Hyde et al., "The Future of Sex and Gender in Psychology"; K. T. F. Kung, K. Louie, D. Spencer, and M. Hines. 2024. "Prenatal Androgen Exposure and Sex-Typical Play Behaviour: A Meta-analysis of Classic Congenital Adrenal Hyperplasia Studies." *Neuroscience and Biobehavioral Reviews* 159: 105616. https://doi.org/10.1016/j.neubiorev.2024.105616.

60. Clancy, *Period.*

61. Clancy, *Period.*

62. Clancy, *Period.*

63. O. J. Fischer. 2021. "Non-binary Reproduction: Stories of Conception, Pregnancy, and Birth." *International Journal of Transgender Health* 22 (1–2): 77–88. https://doi.org/10.1080/26895269.2020.1838392.

64. J. M. Kepley, K. Bates, and S. S. Mohiuddin. 2023. "Physiology, Maternal Changes." StatPearls [Internet], March 12. https://www.ncbi.nlm.nih.gov/books/NBK539766/.

65. M. Blau, R. Hazani, and D. Hekmat. 2016. "Anatomy of the Gynecomastia Tissue and Its Clinical Significance." *Plastic Reconstruction Surgery—Global Open* 4 (8): e854. https://doi.org/10.1097/GOX.0000000000000844.

66. K. Voon and B. G. A. Stuckey. 2023. "Induction of Lactation in a Patient with Complete Androgen Insensitivity Syndrome." *Endocrinology, Diabetes, and Metabolism Case Reports* 4: 23–0063. https://doi.org/10.1530/EDM-23-0063.

67. T. H. Kunz and D. J. Hosken. 2009. "Male Lactation: Why, Why Not and Is It Care?" *Trends in Ecology and Evolution* 24 (2): 80–85. https://doi.org/10.1016/j.tree.2008.09.009.

68. L. Eliot, A. Ahmed, H. Khan, and J. Patel. 2021. "Dump the 'Dimorphism': Comprehensive Synthesis of Human Brain Studies Reveals Few Male-Female Differences beyond Size." *Neurosciences Biobehavioral Reviews* 125: 667–97. https://doi.org/10.1016/j.neubiorev.2021.02.026.

69. G. Rippon. 2019. *Gendered Brain: The New Neuroscience That Shatters the Myth of the Female Brain.* London: The Bodley Head Ltd.; D. Joel, Z. Berman, I. Tavor, N. Wexler, O. Gaber, Y. Stein, N. Shefi, et al. 2015. "Sex Beyond the Genitalia: The Human Brain Mosaic." *Proceedings of the National Academy of Sciences of the United*

States of America 112 (50): 15468–73; D. Joel. 2012. "Genetic-Gonadal-Genitals Sex (3G-Sex) and the Misconception of Brain and Gender, or, Why 3G-Males and 3G-Females Have Intersex Brain and Intersex Gender." *Biology of Sex Differences* 3 (1): 27; S. Liu, J. Seidlitz, J. D. Blumenthal, L. S. Clasen, and A. Raznahan. 2020. "Integrative Structural, Functional, and Transcriptomic Analyses of Sex-Biased Brain Organization in Humans." *Proceedings of the National Academy of Sciences of the United States of America* 117 (31): 18788–98. https://doi.org/10.1073/pnas.1919091117; M. Hirnstein, K. Hugdahl, and M. Hausmann. 2019. "Cognitive Sex Differences and Hemispheric Asymmetry: A Critical Review of 40 Years of Research." *Laterality: Asymmetries of Body, Brain and Cognition* 24 (2): 204–52. https://doi.org/10.1080/1357650x.2018.1497044.

70. Daphna et al., "Sex Beyond the Genitalia," 15468.

71. The nonlinear scaling relationship between a region and brain size (e.g., total brain volume).

72. Called Freesurfer: https://surfer.nmr.mgh.harvard.edu/.

73. C. M. Williams, H. Peyre, R. Toro, and F. Ramus. 2021. "Neuroanatomical Norms in the UK Biobank: The Impact of Allometric Scaling, Sex, and Age." *Human Brain Mapping* 42 (14): 4623–42. https://doi.org/10.1002/hbm.25572.

74. A. R. DeCasien, E. Guma, S. Liu, and A. Raznahan. 2022. "Sex Differences in the Human Brain: A Roadmap for More Careful Analysis and Interpretation of a Biological Reality." *Biological Sex Differences* 13 (1): 43. https://doi.org/10.1186/s13293-022-00448-w.

75. S. J. Ritchie, S. R. Cox, X. Shen, M. V. Lombardo, L. M. Reus, C. Alloza, M. A. Harris, et al. 2018. "Sex Differences in the Adult Human Brain: Evidence from 5216 UK Biobank Participants." *Cerebral Cortex* 28 (8): 2959–75. https://doi.org/10.1093/cercor/bhy109.

76. L. Eliot. 2024. "Remembering the Null Hypothesis When Searching for Brain Sex Differences." *Biology of Sex Differences* 15:14. https://doi.org/10.1186/s13293-024-00585-4.

77. M. Ingalhalikar, A. Smith, D. Parker, T. D. Satterthwaite, M. A. Elliott, K. Ruparel, H. Hakonarson, et al. 2014. "Sex Differences in the Structural Connectome of the Human Brain." *Proceedings of the National Academy of Sciences of the United States of America* 111 (2): 823–28. https://doi.org/10.1073/pnas.1316909110; J. M. Rauch and L. Eliot. 2022. "Breaking the Binary: Gender versus Sex Analysis in Human Brain Imaging." *Neuroimage* 264: 119732. https://doi.org/10.1016/j.neuroimage.2022.119732; Rippon, *Gendered Brain*; R. Jordan-Young. 2011. *Brainstorm: The Flaws in the Science of Brain Differences*. Cambridge, MA: Harvard University Press.

78. G. Downey and D. H. Lende. 2012. "Evolution and the Brain." In *The Encultured Brain: An Introduction to Neuroanthropology*, edited by D. H. Lende and G. Downey, 103–38. Cambridge, MA: MIT Press.

79. Rauch and Eliot, "Breaking the Binary"; Eliot, "Remembering the Null Hypothesis"; Rippon, *Gendered Brain*; L. Eliot, A. K. Beery, E. G. Jacobs, H. F. LeBlanc, D. L. Maney, and M. M. McCarthy. 2023. "Why and How to Account for Sex and Gender in Brain and Behavioral Research." *Journal of Neuroscience* 143 (37): 6344–56. https://doi.org/10.1523/JNEUROSCI.0020-23.2023; A. Kaiser. 2012. "Re-Conceptualizing 'Sex' and 'Gender' in the Human Brain." *Zeitschrift für Psychologie* 220: 130–36; D. L. Maney. 2015. "Just Like a Circus: The Public Consumption of Sex Differences." *Current Topics in Behavioral Neuroscience* 19: 279–96. https://doi.org/10.1007/7854_2014_339.

80. DeCasien et al., "Sex Differences in the Human Brain."

81. Eliot et al., "Dump the 'Dimorphism'"; Eliot, "Remembering the Null Hypothesis."

82. S. Richardson. 2013. *Sex Itself: The Search for Male and Female in the Human Genome*. Chicago: The University of Chicago Press; Kung et al., "Prenatal Androgen Exposure and Sex-Typical Play Behaviour."

83. Many of the genes on the X chromosome are not related directly or only to reproductive biology, but most of the genes on the Y are.

84. Prum, *Performance All the Way Down*.

85. B. Croft, T. Ohnesorg, J. Hewitt, J. Bowles, A. Quinn, J. Tan, V. Corbin, et al. 2018. "Human Sex Reversal Is Caused by Duplication or Deletion of Core Enhancers Upstream of SOX9." *Nature Communications* 9 (1): 5319. https://doi.org/10.1038/s41467-018-07784-9; P. Bernard and V. R. Harley. 2007. "Wnt4 Action in Gonadal Development and Sex Determination." *The International Journal of Biochemistry & Cell Biology* 39 (1): 31–43. https://doi.org/10.1016/j.biocel.2006.06.007; K. K. Niakan and E. R. B. McCabe. 2005. "DAX1 Origin, Function, and Novel Role." *Molecular Genetics and Metabolism* 86 (1): 70–83. https://doi.org/10.1016/j.ymgme.2005.07.019.

86. Fausto-Sterling, *Sexing the Body*; Prum, *Performance All the Way Down*.

87. O. M. Muñoz-Aguirre, S. Kim-Hellmuth, V. Wucher, A. D. H. Gewirtz, D. J. Cotter, P. Parsana, S. Kasela, et al. 2020. "The Impact of Sex on Gene Expression across Human Tissues." *Science* 369 (6509): eaba3066. https://doi.org/10.1126/science.aba3066.

88. M. Gershoni and S. Pietrokovski. 2017. "The Landscape of Sex-Differential Transcriptome and Its Consequent Selection in Human Adults." *BMC Biol* 15 (7). https://doi.org/10.1186/s12915-017-0352-z.

89. Richardson, *Sex Itself*.

90. A. Wiley. 2021. "Pearl Lecture: Biological Normalcy: A New Framework for Biocultural Analysis of Human Population Variation." *American Journal of Human Biology* 33 (5): e23563. https://doi.org/10.1002/ajhb.23563.

91. See Fausto-Sterling, *Sexing the Body*, for an excellent and extensive overview of these patterns and their explanations and impacts. See also C. Ainsworth. 2015. "Sex Redefined: The Idea of 2 Sexes Is Overly Simplistic." *Scientific American*,

February 8. https://www.scientificamerican.com/article/sex-redefined-the-idea-of-2-sexes-is-overly-simplistic1; and Richardson, *Sex Itself.*

92. Hyde et al., "The Future of Sex and Gender in Psychology"; and Fausto-Sterling, *Sexing the Body*; S. M. van Anders. 2022. "Gender/Sex/ual Diversity and Biobehavioral Research." *Psychology of Sexual Orientation and Gender Diversity*, November 10. Advance online publication. https://dx.doi.org/10.1037/sgd0000609.

6. No Biological Battle of the Sexes

1. This book has sold more than fifteen million copies, been reprinted many times, and remains a near best seller in the 2020s. John Gray's website offers (for a fee) to help you know who your soul mate is, increase intimacy and passion, create romance with your partner, and enjoy a lifetime of great sex (Marsvenus.com).

2. C. Darwin. 1871. *The Descent of Man, and Selection in Relation to Sex.* London: John Murray, 401.

3. A. J. Bateman. 1948. "Intra-sexual Selection in Drosophila." *Heredity* 2: 349–68; E. O. Wilson. 1979. "Sex and Human Nature." *The Wilson Quarterly* 3 (4): 92–105.

4. H. Dunsworth. 2021. "This View of Wife." In *A Most Interesting Problem*, edited by J. DaSilva, chap. 9. Princeton, NJ: Princeton University Press; L. Cooke. 2021. *Bitch: On the Female of the Species.* New York: Basic Books; M. Landau. 1991. *Narrative of Human Evolution.* New Haven, CT: Yale University Press; A. Fausto-Sterling. 1992. *Myths of Gender.* New York: Basic Books; M. Midgley. 2003. *The Myths We Live By.* Oxfordshire: Routledge; A. Saini. 2017. *Inferior: How Science Got Women Wrong—and the New Research That's Rewriting the Story.* Boston: Beacon Press.

5. W. D. Lassek and S. J. C. Gaulin. 2022. "Substantial but Misunderstood Human Sexual Dimorphism Results Mainly from Sexual Selection on Males and Natural Selection on Females." *Frontiers in Psychology* 13: 859931. https://doi.org/10.3389/fpsyg.2022.859931. These authors state specifically that (1) sexual dimorphism in stature, fat mass, and fat distribution have been significantly shaped by disruptive natural-selection regimes operating on females and males, with some likely overlay of subsequent sexual selection acting via mate choice in the case of fat distribution; (2) sexual dimorphism in lean mass, muscle mass, and strength are largely due to sexual selection arising from a long history of aggressive male mating competition, with the some possible influence of divergent natural selection due to sex differences in foraging ecology; and (3) a large literature seems to have overemphasized the role of mate choice, and underestimated the role of male competition for mates, in shaping human sex differences. For a version of this same argument that focuses on the female need to stay healthy to raise young, see here: J. F. Benenson, C. E. Webb, and R. W. Wrangham. 2021. "Self-Protection as an Adaptive Female Strategy." *Behavioral and Brain Sciences* 45: e128. https://doi.org/10.1017/S0140525X21002417.

6. See C. Hooven. 2021. *T: The Story of Testosterone, the Hormone That Dominates and Divides Us.* New York: Henry Holt and Company, 25.

7. See, for example, R. Jordan-Young and K. Karkazis. 2019. *Testosterone: An Unauthorized Biography.* Cambridge, MA: Harvard University Press; C. Fine. 2016. *Testosterone Rex: Myths of Sex, Science and Society.* New York: W. W. Norton; S. M. van Anders. 2013. "Beyond Masculinity: Testosterone, Gender/Sex, and Human Social Behavior in a Comparative Context." *Frontiers in Neuroendocrinology* 34: 198–210. https://doi.org/10.1016/j.yfrne.2013.07.001; S. M. van Anders, J. Steiger, and K. L. Goldey. 2015. "Effects of Gendered Behavior on Testosterone in Women and Men." *Proceedings of the National Academy of Sciences of the United States of America* 112: 13805–10; A. Fausto-Sterling. 2020. *Sexing the Body: Gender Politics and the Construction of Sexuality.* London: Hachette; S. Richardson. 2013. *Sex Itself: The Search for Male and Female in the Human Genome.* Chicago: The University of Chicago Press; L. T. Gettler. 2016. "Becoming DADS: Considering the Role of Cultural Context and Developmental Plasticity for Paternal Socioendocrinology." *Current Anthropology* 57: S38–S51.

8. Meta-analyses are systematic syntheses of many studies.

9. As usual, these studies are mostly on people in North America and Europe. Thus, they are probably underrepresenting the broader variation in humans across the planet.

10. While the authors use the terms "male" and "female," they are actually assessing gender (men/women, boys/girls) in these studies. J. Shibley Hyde. 2005. "The Gender Similarities Hypothesis." *American Psychologist* 60 (6): 581–92; E. Zell, Z. Krizan, and S. R. Teeter. 2015. "Evaluating Gender Similarities and Differences Using Metasynthesis." *American Psychologist* 70: 10–20. https://doi.org/10.1037/a0038208.

11. Zell, Krizan, and Teeter, "Evaluating Gender Similarities and Differences."

12. R. W. Wrangham and M. L. Wilson. 2004. "Collective Violence—Comparisons between Youths and Chimpanzees." In *Youth Violence: Scientific Approaches to Prevention*, edited by J. Devine, J. Gilligan, K. A. Miczek, R. Shaikh, and D. Pfaff, 233–56. New York: New York Academy of Sciences; Hooven, *T: The Story of Testosterone*, 25; Benenson, Webb, and Wrangham, "Self-Protection as an Adaptive Female Strategy"; Lassek and Gaulin, "Substantial but Misunderstood Human Sexual Dimorphism"; D. Buss and D. P. Schmitt. 2019. "Mate Preferences and Their Behavioral Manifestations." *Annual Review of Psychology* 70: 77–110. See also previous mention of Darwin, E. O. Wilson, Robert Trivers, etc. (chapter 3). See also R. Nelson. 2021. "The Sex in Your Violence: Patriarchy and Power in Anthropological World Building and Everyday Life." *Current Anthropology* 62: S92–S102. https://doi.org/10.1086/711605.

13. M. G. Alexander and T. D. Fisher. 2002. "Truth and Consequences: Using the Bogus Pipeline to Examine Sex Differences in Self-Reported Sexuality." *Journal of Sex Research* 40 (1): 27–35; H. Donnan and F. Magowan. 2010. *The Anthropology of Sex.* Oxfordshire: Routledge.

14. R. Thomsen. 2013. "Masturbation in Non-Human Primates." In *The Encyclopedia of Human Sexuality*, vol. 1. Boston: Wiley-Blackwell Press.

15. Fausto-Sterling, *Sexing the Body*.

16. The statistics are summarized data presented in the categories used by the main data article: D. Herbenick, M. Reece, V. Schick, S. A. Sanders, B. Dodge, and J. D. Fortenberry. 2010. "Sexual Behavior in the United States: Results from a National Probability Sample of Men and Women Ages 14–94." *Journal of Sexual Medicine* 7 (suppl. 5): 255–65.

17. The authors of the study make the point that their survey undersampled homosexually oriented populations.

18. C. Ryan and C. Jetha. 2010. *Sex at Dawn: The Prehistoric Origins of Modern Sexuality*. New York: Harper; Donnan and Magowan, *Anthropology of Sex*; J. L. Carroll. 2009. *Sexuality Now: Embracing Diversity*, 3rd ed. Boston: Cenage Learning; and also J. DeLamater and W. N. Friedrich. 2002. "Human Sexual Development." *Journal of Sex Research* 39 (1): 10–14.

19. See S. T. Lindau, L. P. Schumm, E. O. Laumann, W. Levinson, C. A. O'Muircheartaigh, and L. J. Waite. 2007. "A Study of Sexuality and Health among Older Adults in the United States." *New England Journal of Medicine* 357 (8): 762–74; and A. Brewis and M. Meyer. 2005. "Marital Coitus across the Life Course." *Journal of Biocultural Science* 37: 499–518.

20. D. P. Schmitt. 2005. "Sociosexuality from Argentina to Zimbabwe: A 48-Nation Study of Sex, Culture, and Strategies of Human Mating." *Behavioral and Brain Sciences* 28: 247–311; but see also the update on this study for more complexity and a less clear set of assertions about mate choice patterns than in the 1993 and 2005 original studies: Buss and Schmitt, "Mate Preferences and Their Behavioral Manifestations."

21. K. Wellings, M. Collumbien, E. Slaymaker, S. Singh, Z. Hodges, D. Patel, and N. Bajos. 2006. "Sexual Behaviour in Context: A Global Perspective." *Lancet* 368 (9548): 1706–28. https://doi.org/10.1016/S0140-6736(06)69479-8.

22. Gettler, "Becoming DADS."

23. M. Cappelletti and K. Wallen. 2016. "Increasing Women's Sexual Desire: The Comparative Effectiveness of Estrogens and Androgens." *Hormones and Behavior* 78: 178–93. https://doi.org/10.1016/j.yhbeh.2015.11.003; S. J. Parish, J. A. Simon, S. R. Davis, A. Giraldi, I. Goldstein, S. W. Goldstein, N. N. Kim, et al. 2021. "International Society for the Study of Women's Sexual Health Clinical Practice Guideline for the Use of Systemic Testosterone for Hypoactive Sexual Desire Disorder in Women." *Climacteric* 24 (6): 533–50. https://doi.org/10.1080/13697137.2021.1891773.

24. S. M. van Anders. 2012. "Testosterone and Sexual Desire in Healthy Women and Men." *Archives of Sexual Behavior* 41 (6): 1471–84. https://doi.org/10.1007/s10508-012-9946-2.

25. Schmitt, "Sociosexuality from Argentina to Zimbabwe," 273.

26. In fact, there is good evidence that gender roles and cultural pressures affect the way people respond to questions about sexuality. See Alexander and Fisher, "Truth and Consequences."

27. W. Wood and A. H. Eagly. 2002. "A Cross-Cultural Analysis of the Behavior of Women and Men: Implications for the Origins of Sex Differences." *Psychological Bulletin* 128 (5): 699–727; and J. Archer. 2009. "Does Sexual Selection Explain Human Sex Differences in Aggression?" *Behavioral and Brain Sciences* 32: 249–311.

28. R. Bribiescas. 2021. "Evolutionary and Life History Insights into Masculinity and Warfare: Opportunities and Limitations." *Current Anthropology* 62: S23.

29. There is little extensive data on violence and homosexual partnerships.

30. J. Archer. 2019. "The Reality and Evolutionary Significance of Psychological Sex Differences." *Biological Reviews* 94 (4): 1381–1415. https://doi.org/10.1111/brv.12507; and Zell, Krizan, and Teeter, "Evaluating Gender Similarities and Differences."

31. M. Gutmann. 2019. *Are Men Animals?* New York: Basic Books; M. Guttman, R. G. Nelson, and A. Fuentes. 2021. "Epidemic Errors in Understanding Masculinity, Maleness, and Violence." *Current Anthropology* 62: S5–S12. https://doi.org/10.1086/712485; A. Fuentes. 2022. *Race, Monogamy and Other Lies They Told You: Busting Myths about Human Nature*. 2nd ed. Oakland: University of California Press.

32. There is a wide array of definitions and framings of sexual assault that vary by and are related to cultural and gendered beliefs, roles, laws, and experiences. This makes global assessments of what percentage of men perpetrate it, and how to accurately measure global patterns complex. However, the WHO definition is a good working frame: "any sexual act, attempt to obtain a sexual act, unwanted sexual comments or advances, or acts to traffic, or otherwise directed, against a person's sexuality using coercion, by any person regardless of their relationship to the victim, in any setting, including but not limited to home and work." See P. A. Fernandez. 2011. "Sexual Assault: An Overview and Implications for Counselling Support." *Australasian Medical Journal* 4 (11): 596–602. https://doi.org/10.4066/AMJ.2011858; and R. Jewkes, C. Garcia-Moren, and P. Sen. 2002. "Sexual Violence." In *World Report on Violence and Health*, edited by E. G. Krug, L. L. Dahlberg, J. A. Mercy, A. B. Zwi, and R. Lozano, 147–74. Geneva: World Health Organization.

33. Nelson, "The Sex in Your Violence"; Fernandez, "Sexual Assault."

34. See R. W. Sussman and C.R. Cloninger. 2011. *Cooperation and Altruism*. New York: Springer; and D. Fry. 2007. *Beyond War: The Human Potential for Peace*. Oxford: Oxford University Press; Fuentes, *Race, Monogamy and Other Lies*; A. Fuentes. 2021. "Searching for the 'Roots' of Masculinity in Primates and the Human Evolutionary Past." *Current Anthropology* 62: S13–S25. https://doi.org/10.1086/711582; Guttman, *Are Men Animals?*

35. G. Rippon. 2019. *Gendered Brain: The New Neuroscience That Shatters the Myth of the Female Brain*. London: The Bodley Head Ltd.; D. Joel, Z. Berman, I. Tavor, N. Wexler, O. Gaber, Y. Stein, N. Shefi, et al. 2015. "Sex Beyond the Genitalia: The Human Brain Mosaic." *Proceedings of the National Academy of Sciences of the United States of America* 112 (50): 15468–73; D. Joel. 2012. "Genetic-Gonadal-Genitals Sex (3G-Sex) and the Misconception of Brain and Gender, or, Why 3G-Males and 3G-Females Have Intersex Brain and Intersex Gender." *Biology of Sex Differences* 3 (1): 27; S. Liu, J. Seidlitz, J. D. Blumenthal, L. S. Clasen, and A. Raznahan. 2020. "Integrative Structural, Functional, and Transcriptomic Analyses of Sex-Biased Brain Organization in Humans." *Proceedings of the National Academy of Sciences of the United States of America* 117 (31): 18788–98. https://doi.org/10.1073/pnas.1919091117; M. Hirnstein, K. Hugdahl, and M. Hausmann. 2019. "Cognitive Sex Differences and Hemispheric Asymmetry: A Critical Review of 40 Years of Research." *Laterality: Asymmetries of Body, Brain and Cognition* 24 (2): 204–52. https://doi.org/10.1080/1357650x.2018.1497044.

36. Joel et al., "Sex Beyond the Genitalia."

37. L. Eliot, A. Ahmed, H. Khan, and J. Patel. 2021. "Dump the 'Dimorphism': Comprehensive Synthesis of Human Brain Studies Reveals Few Male-Female Differences beyond Size." *Neuroscience and Biobehavioral Reviews* 125: 667–97. https://doi.org/10.1016/j.neubiorev.2021.02.026; L. Eliot. 2024. "Remembering the Null Hypothesis When Searching for Brain Sex Differences." *Biology of Sex Differences* 15:14. https://doi.org/10.1186/s13293-024-00585-4.

38. L. S. B. Hrdy. 2009. *Mother and Others: The Evolutionary Origins of Mutual Understanding*. New York: Belknap; C. Kuzawa, H. T. Chugani, L. I. Grossman, L. Lipovich, O. Muzik, P. Hof, R. Patrick, D. E. Wildman, et al. 2014. "Metabolic Costs and Evolutionary Implications of Human Brain Development." *Proceedings of the National Academy of Sciences of the United States of America* 111 (36): 13010–15; K. R. Rosenberg. 2021. "The Evolution of Human Infancy: Why It Helps to Be Helpless." *Annual Review of Anthropology* 50: 423–40. https://doi.org/10.1146/annurev-anthro-111819-105454.

39. K. L. Bales, C. S. Ardekani, A. Baxter, C. L. Karaskiewicz, J. X. Kuske, A. R. Lau, L. E. Savidge, et al. 2021. "What Is a Pair Bond?" *Hormones and Behavior* 36: 105062. https://doi.org/10.1016/j.yhbeh.2021.105062; A. Fuentes. 1999. "Re-evaluating Primate Monogamy." *American Anthropologist* 100 (4): 890–907; A. Fuentes. 2002. "Patterns and Trends in Primate Pair Bonds." *International Journal of Primatology* 23 (4): 953–78.

40. See A. Fuentes. 2017. *The Creative Spark: How Imagination Made Humans Exceptional*. New York: Dutton; but for a counterargument, see B. Chapais. 2010. *Primeval Kinship: How Pair-Bonding Gave Birth to Human Society*. Cambridge, MA: Harvard University Press.

41. J. S. Hyde, R. S. Bigler, D. Joel, C. C. Tate, and S. M. van Anders. 2019. "The Future of Sex and Gender in Psychology: Five Challenges to the Gender Binary." *American Psychologist* 74 (2): 15. https://doi.org/10.1037/amp0000307.

7. Why the Binary View Is a Problem

1. K. O. Smiley, K. M. Munley, K. Aghi, S. E. Lipshutz, T. M. Patton, D. S. Pradhan, T. K. Solomon-Lane, et al. 2024. "Sex Diversity in the 21st Century: Concepts, Frameworks, and Approaches for the Future of Neuroendocrinology." *Hormones and Behavior* 157: 105445. https://doi.org/10.1016/j.yhbeh.2023.105445.

2. S. M. van Anders. 2013. "Beyond Masculinity: Testosterone, Gender/Sex, and Human Social Behavior in a Comparative Context." *Frontiers in Neuroendocrinology* 34: 198–210. https://doi.org/10.1016/j.yfrne.2013.07.001; A. Fausto-Sterling. 2020. *Sexing the Body: Gender Politics and the Construction of Sexuality*. London: Hachette; S. Richardson. 2013. *Sex Itself: The Search for Male and Female in the Human Genome.* Chicago: The University of Chicago Press.

3. S. LeVay. 1991. "A Difference in Hypothalamic Structure between Heterosexual and Homosexual Men." *Science* 253 (5023): 1034–37. https://doi.org/10.1126/science.1887219.

4. "Researcher Simon LeVay Talks About the Science of Sexuality." Elmhurst University, October 18, 2012. https://www.elmhurst.edu/news/researcher-simon-levay-talks-about-the-science-of-sexuality/.

5. A. Borsa, M. Miyagi, K. Ichikawa, K. De Leon De Jesus, K. Jillson, M. Boulicault, and S. S. Richardson. 2024. "The New Genetics of Sexuality." *GLQ* 30 (1): 119–40; A. Ganna, K. J. H. Verweij, M. G. Nivard, R. Maier, R. Wedow, A. S. Busch, A. Abdellaoui, et al. 2019. "Large-Scale GWAS Reveals Insights into the Genetic Architecture of Same-Sex Sexual Behavior." *Science* 365 (6456): eaat7693. https://doi.org/10.1126/science.aat7693; Fausto-Sterling, *Sexing the Body*; C. E. Roselli. 2018. "Neurobiology of Gender Identity and Sexual Orientation." *Journal of Neuroendocrinology* 30 (7): e12562. https://doi.org/10.1111/jne.12562.

6. Fausto-Sterling, *Sexing the Body*; Roselli, "Neurobiology of Gender Identity and Sexual Orientation"; A. Fuentes. 2022. *Race, Monogamy and Other Lies They Told You: Busting Myths about Human Nature*. 2nd ed. Oakland: University of California Press; C. Hooven. 2021. *T: The Story of Testosterone, the Hormone That Dominates and Divides Us.* New York: Henry Holt and Company; R. Jordan-Young and K. Karkazis. 2019. *Testosterone: An Unauthorized Biography*. Cambridge, MA: Harvard University Press; C. Fine. 2016. *Testosterone Rex: Myths of Sex, Science and Society*. New York: W. W. Norton; van Anders, "Beyond Masculinity"; M. Gutmann. 2019. *Are Men Animals?* New York: Basic Books; and many, many more. . . .

7. Once the initial sensory stimulus engages the limbic system, neural stimuli (brain actions) are transmitted to much of the body via the endocrine system (using the same hormones, testosterone, vasopressin, and oxytocin, in all humans). This leads to physical and psychological excitement that includes increased blood flow and swelling of tissues (called vasocongestion) and a tensing of the muscles (called myotonia) throughout the body, sporadic increases in blood pressure, lubrication in the vagina and inner labia, erection of the clitoris and erection of the penis, glandular secretions across parts of the body, and ultimately (or potentially) a variable suite of physiological changes associated with orgasm. Fuentes, *Race, Monogamy and Other Lies*; see also R. S. Calabrò, A. Cacciola, D. Bruschetta, et al. 2010. "Neuroanatomy and Function of Human Sexual Behavior: A Neglected or Unknown Issue?" *Brain and Behavior* 9: e01389. https://doi.org/10.1002/brb3.1389.

8. See S. T. Lindau, L. P. Schumm, E. O. Laumann, W. Levinson, C. A. O'Muircheartaigh, and L. J. Waite. 2007. "A Study of Sexuality and Health among Older Adults in the United States." *New England Journal of Medicine* 357 (8): 762–74; and Brewis and Meyer, "Marital Coitus across the Life Course."

9. C. M. Curley and B. T. Johnson 2022. "Sexuality and Aging: Is It Time for a New Sexual Revolution?" *Social Science and Medicine* 301: 114865. https://doi.org/10.1016/j.socscimed.2022.114865; see also "Sexuality and Intimacy in Older Adults." NIH National Institute on Aging, accessed September 8, 2024. https://www.nia.nih.gov/health/sexuality/sexuality-and-intimacy-older-adults.

10. Associated Press. 2020. "Tennessee Governor Signs Anti-Gay Adoption Bill." NBC News, January 24. https://www.nbcnews.com/feature/nbc-out/tennessee-governor-signs-anti-gay-adoption-bill-n1122436.

11. "Child Welfare Nondiscrimination Laws." Movement Advancement Project, accessed September 8, 2024. https://www.lgbtmap.org/equality-maps/foster_and_adoption_laws.

12. "Same-Sex Parents Are 7 Times More Likely to Raise Adopted and Foster Children." UCLA Williams Institute, accessed September 8, 2024. https://williamsinstitute.law.ucla.edu/press/lgbt-parenting-media-alert/.

13. R. Sear. 2021. "The Male Breadwinner Nuclear Family Is Not the 'Traditional' Human Family, and Promotion of This Myth May Have Adverse Health Consequences." *Philosophical Transactions of the Royal Society: Biological Sciences* 376: 20200020. https://doi.org/10.1098/rstb.2020.0020; E. Hubbard, O. Shannon, and A. Pisor. 2023. "Non-kin Alloparents and Child Outcomes: Older Siblings, but Not Godparents, Predict Educational Attainment in a Rural Context." *Evolution and Human Behavior* 44: 597–604. https://doi.org/10.1016/j.evolhumbehav.2023.09.006; K. L. Kramer. 2021. "The Human Family—Its Evolutionary Context and Diversity." *Social Sciences* 10 (6): 191. https://doi.org/10.3390/socsci10060191.

14. There are also many other, very serious, problems with animal testing. See H. Ferdowsian, A. Fuentes, L. Johnson, B. King, and J. Pierce. 2022. "Toward an Anti-Maleficent Research Agenda." *Cambridge Quarterly of Healthcare Ethics* 31 (1): 54–58. https://doi.org/10.1017/S0963180121000487; and B. Keim. 2023. "What Do We Owe Lab Animals?" *New York Times*, January 23. https://www.nytimes.com/2023/01/23/science/what-do-we-owe-lab-animals.html.

15. This is not true of all medical researchers, especially in the past decade or two. Today, there are many outspoken advocates of including female nonhuman animals (and humans) in research. Most major government funding agencies of Canada, the United States, and the European Union now require such inclusion.

16. For an excellent overview, see E. Cleghorn. 2021. *Unwell Women: Misdiagnosis and Myth in a Man-Made World*. New York: Dutton.

17. A. Ghorayshi. 2023. "Guess Which Sex Behaves More Erratically (at Least in Mice)." *New York Times*, March 7. https://www.nytimes.com/2023/03/07/science/female-mice-hormones.html.

18. Which was primarily the United States and Europe until the last fifty years or so.

19. There is a range of solid scholarship on this blend of racism and sexism and how it shaped contemporary medical practice in the United States; see Cleghorn, *Unwell Women*; C. Willoughby. 2022. *Masters of Health: Racial Science and Slavery in US Medical Schools*. University of North Carolina Press. See also A. K. Beery and I. Zucker. 2011. "Sex Bias in Neuroscience and Biomedical Research." *Neuroscience and Biobehavioral Reviews* 35 (3): 565–72. https://doi.org/10.1016/j.neubiorev.2010.07.002; Ghorayshi, "Guess Which Sex Behaves More Erratically"; D. R. Levy, N. Hunter, S. Lin, E. M. Robinson, W. Gillis, E. B. Conlin, R. Anyoha, et al. 2023. "Mouse Spontaneous Behavior Reflects Individual Variation Rather Than Estrous State." *Current Biology* 33: 1–7.

20. L. Greaves and S. A. Ritz. 2022. "Sex, Gender and Health: Mapping the Landscape of Research and Policy." *International Journal of Environmental Research and Public Health* 19: 2563. https://doi.org/10.3390/ijerph19052563.

21. This is primarily referring to 3G females, but we know that the term "women" is not a biological term and not all women are 3G females. This, however, still supports the statement as the range of biological variation in the term "women" (and in "men") is much larger than is envisioned, or engaged, by much medical research.

22. I. Zucker and B. J. Prendergast. 2020. "Sex Differences in Pharmacokinetics Predict Adverse Drug Reactions in Women." *Biology of Sex Differences* 11: 32. https://doi.org/10.1186/s13293-020-00308-5.

23. If this is the case, there are not really any significant differences between male and female ADEs. T. Rushovich, A. Gompers, J. W. Lockhart, et al. 2023. "Adverse Drug Events by Sex after Adjusting for Baseline Rates of Drug Use." *JAMA Netw Open*. 6 (8): e2329074. https://doi.org/10.1001/jamanetworkopen.2023.29074.

24. D. J. Greenblatt, J. S. Harmatz, and T. Roth. 2019. "Zolpidem and Gender: Are Women Really at Risk?" *Journal of Clinical Psychopharmacology* 39 (3): 189–99. https://doi.org/10.1097/JCP.0000000000001026. PMID: 30939589.

25. H. Zhao, M. DiMarco, K. Ichikawa, M. Boulicault, M. Perret, K. Jillson, A. Fair, et al. 2023. "Making a 'Sex-Difference Fact': Ambien Dosing at the Interface of Policy, Regulation, Women's Health, and Biology." *Social Studies of Science* 53 (4): 475–94. https://doi.org/10.1177/03063127231168371.

26. M. Roser, C. Appel, and H. Ritchie. 2013. "Human Height." OurWorldInData.org. Last updated January 2024. https://ourworldindata.org/human-height; T. G. Travison, H. W. Vesper, E. Orwoll, F. Wu, J. M. Kaufman, Y. Wang, B. Lapauw, et al. 2017. "Harmonized Reference Ranges for Circulating Testosterone Levels in Men of Four Cohort Studies in the United States and Europe." *The Journal of Clinical Endocrinology & Metabolism* 102 (4): 1161–73. https://doi.org/10.1210/jc.2016-2935; I. Janssen, S. B. Heymsfield, Z. Wang, and R. Ross. 2000. "Skeletal Muscle Mass and Distribution in 468 Men and Women aged 18–88 Yr." *Journal of Applied Physiology* 89 (1): 81–88.

27. Roser, Appel, and Ritchie, "Human Height."

28. See references for humans in V. Puppo. 2013. "Anatomy and Physiology of the Clitoris, Vestibular Bulbs, and Labia Minora with a Review of the Female Orgasm and the Prevention of Female Sexual Dysfunction." *Clinical Anatomy* 26: 134–52.

29. D. E. McGhee and J. R. Steele. 2020. "Breast Biomechanics: What Do We Really Know?" *Physiology* 35 (2): 144–56.

30. K. Clancy. 2023. *Period: The Real Story of Menstruation*. Princeton, NJ: Princeton University Press.

31. L. Eliot, A. Ahmed, H. Khan, and J. Patel. 2021. "Dump the 'Dimorphism': Comprehensive Synthesis of Human Brain Studies Reveals Few Male-Female Differences beyond Size." *Neuroscience Biobehavioral Reviews* 125: 667–97. https://doi.org/10.1016/j.neubiorev.2021.02.026; L. Eliot. 2024. "Remembering the Null Hypothesis When Searching for Brain Sex Differences." *Biology of Sex Differences* 15:14. https://doi.org/10.1186/s13293-024-00585-4.

32. K. Sobhani, D. K. Nieves Castro, Q. Fu, R. A. Gottlieb, J. E. Van Eyk, and C. Noel Bairey Merz. 2018. "Sex Differences in Ischemic Heart Disease and Heart Failure Biomarkers." *Biology of Sex Differences* 9 (1): 43. https://doi.org/10.1186/s13293-018-0201-y.

33. L. J. Shaw, R. Bugiardini, and C. N. Merz. 2009. "Women and Ischemic Heart Disease: Evolving Knowledge." *Journal of the American College of Cardiologists* 54 (17): 1561–75. https://doi.org/10.1016/j.jacc.2009.04.098.

34. A. Gasecka, J. M. Zimodro, and Y. Appelman. 2023. "Sex Differences in Antiplatelet Therapy: State-of-the Art." *Platelets* 34 (1): 2176173. https://doi.org/10.1080/09537104.2023.2176173.

35. W. M. Schultz, H. M. Kelli, J. C. Lisko, T. Varghese, J. Shen, P. Sandesara, A. A. Quyyumi, et al. 2018. "Socioeconomic Status and Cardiovascular Outcomes: Challenges and Interventions." *Circulation* 137 (20): 2166–78. https://doi.org/10.1161/CIRCULATIONAHA.117.029652; T. M. Powell-Wiley, Y. Baumer, F. O. Baah, A. S. Baez, N. Farmer, C. T. Mahlobo, M. A. Pita, et al. 2022. "Social Determinants of Cardiovascular Disease." *Circulation Research* 130 (5): 782–99. https://doi.org/10.1161/CIRCRESAHA.121.319811.

36. See Brigham and Woman's Hospital information as a good example: "Heart Disease: 7 Differences between Men and Women." Accessed September 8, 2024. https://give.brighamandwomens.org/7-differences-between-men-and-women/.

37. See F. Puoti, A. Ricci, A. Nanni-Costa, W. Ricciardi, W. Malorni, and E. Ortona. 2016. "Organ Transplantation and Gender Differences: A Paradigmatic Example of Intertwining between Biological and Sociocultural Determinants." *Biological Sex Differences* 7: 35. https://doi.org/10.1186/s13293-016-0088-4.

38. A. Melk, B. Babitsch, B. Borchert-Mörlins, F. Claas, A. I. Dipchand, S. Eifert, B. Eiz-Vesper, et al. 2019. "Equally Interchangeable? How Sex and Gender Affect Transplantation." *Transplantation* 103 (6): 1094–110. https://doi.org/10.1097/TP.0000000000002655.

39. National Academies of Sciences, Engineering, and Medicine. 2020. *Birth Settings in America: Outcomes, Quality, Access, and Choice.* Washington, DC: The National Academies Press. https://nap.nationalacademies.org/resource/25636/interactive/.

40. S. Vedam, K. Stoll, T. K. Taiwo, N. Rubashkin, M. Cheyney, N. Strauss, M. McLemore, et al. 2019. "The Giving Voice to Mothers Study: Inequity and Mistreatment during Pregnancy and Childbirth in the United States." *Reproductive Health* 16: 77. https://doi.org/10.1186/s12978-019-0729-2; L. Hill, S. Artiga, and U. Ranji. 2022. "Racial Disparities in Maternal and Infant Health: Current Status and Efforts to Address Them." KFF, November 1. https://www.kff.org/racial-equity-and-health-policy/issue-brief/racial-disparities-in-maternal-and-infant-health-current-status-and-efforts-to-address-them/.

41. National Academies of Sciences, Engineering, and Medicine, *Birth Settings in America.*

42. A. M. Jukic, D. D. Baird, C. R. Weinberg, D. R. McConnaughey, and A. J. Wilcox. 2013. "Length of Human Pregnancy and Contributors to Its Natural Variation." *Human Reproduction* 28 (10): 2848–55. https://doi.org/10.1093/humrep/det297; A. J. Butwick, M. L. Druzin, G. M. Shaw, and N. Guo. 2020. "Evaluation of US State-Level Variation in Hypertensive Disorders of Pregnancy." *JAMA Network Open* 3 (10): e2018741. https://doi.org/10.1001/jamanetworkopen.2020.18741.

43. Such as training and running into the third trimester and across lactations: A. S. Tenforde, K. E. Toth, E. Langen, M. Fredericson, and K. L. Sainani. 2015.

"Running Habits of Competitive Runners during Pregnancy and Breastfeeding." *Sports Health* 7 (2): 172–76. https://doi.org/10.1177/1941738114549542.

44. The term "sex contextualism" was coined by the philosopher and historian of science Sarah Richardson: see S. S. Richardson. 2022. "Sex Contextualism." *Philosophy, Theory, and Practice in Biology* 14: 2. https://doi.org/10.3998/ptpbio.2096.

45. J. S. Williams, M. R. Fattori, I. R. Honeyborne, and S. A. Ritz. 2023. "Considering Hormones as Sex- and Gender-Related Factors in Biomedical Research: Challenging False Dichotomies and Embracing Complexity." *Hormones and Behavior* 156: 105442. https://doi.org/10.1016/j.yhbeh.2023.105442; M. Pape, M. Miyagi, S. A. Ritz, M. Boulicault, S. S. Richardson, and D. L. Maney. 2024. "Sex Contextualism in Laboratory Research: Enhancing Rigor and Precision in the Study of Sex-Related Variables." *Cell* 187 (6): 1316–26. https://doi.org/10.1016/j.cell.2024.02.008.

46. Pape et al., "Sex Contextualism in Laboratory Research."

47. For a great discussion on this topic, see interview of Sarah Richardson by Heather Shattuck-Heidorn and Kelsey Ichikawa on the GenderSci Lab website: "No Sex without Context: A Q&A with Sarah Richardson on 'Sex Contextualism.'." https://www.genderscilab.org/blog/q-and-a-sarah-richardson-on-sex-contextualism. See also L. Greaves and S. A. Ritz. 2022. "Sex, Gender and Health: Mapping the Landscape of Research and Policy." *International Journal of Environmental Research and Public Health* 19: 2563. https://doi.org/10.3390/ijerph19052563.

48. C. Semenya. 2023. "Running in a Body That's My Own." *The New York Times*, October 21. https://www.nytimes.com/2023/10/21/opinion/running-body-semenya.html.

49. Despite these data and facts, in 2023 World Athletics banned the women with higher natural testosterone levels from all track events unless they underwent six months of hormone-repressing treatment. E. Pells. 2023. "Track Bans Transgender Athletes, Tightens Rules for Semenya." *AP News*, March 23. https://apnews.com/article/transgender-track-semenya-f3499b00b932948f96838adb3b010f11.

50. S. Bermon and P. Y. Garnier. 2017. "Serum Androgen Levels and Their Relation to Performance in Track and Field: Mass Spectrometry Results from 2127 Observations in Male and Female Elite Athletes." *British Journal of Sports Medicine* 51 (17): 1309–14. https://doi.org/10.1136/bjsports-2017-097792.

51. "Correction: Serum Androgen Levels and Their Relation to Performance in Track and Field: Mass Spectrometry Results from 2127 Observations in Male and Female Elite Athletes." *British Journal of Sports Medicine* 2021 55 (17): e7. https://doi.org/10.1136/bjsports-2017-097792corr1. Erratum for: *British Journal of Sports Medicine* 2017 51 (17): 1309–14.

52. D. J. Handelsman, A. L. Hirschberg, and S. Bemon. 2018. "Circulating Testosterone as the Hormonal Basis of Sex Differences in Athletic Performance." *Endocrine*

Reviews 39 (5): 803–29. https://doi.org/10.1210/er.2018-00020. See also Jordan-Young and Karkazis, *Testosterone*.

53. But see S. Bekker and S. Mumford. Forthcoming. *Open Play: The Case for Feminist Sport*. London: Reaktion Books, for a much more in depth and complicated history of why "women's" sports came to be.

54. Bekker and Mumford, *Open Play*; J. L. Parsons, S. E. Coen, and S. Bekker. 2021. "Anterior Cruciate Ligament Injury: Towards a Gendered Environmental Approach." *British Journal of Sports Medicine* 55: 984–90; L. C. Hallam and F. T. Amorim. 2022. "Expanding the Gap: An Updated Look into Sex Differences in Running Performance." *Frontiers in Physiology* 12: 804149. https://doi.org/10.3389/fphys.2021.804149.

55. Hallam and Amorim, "Expanding the Gap."

56. Julian Savulescu. 2019. "Ten Ethical Flaws in the Caster Semenya Decision on Intersex in Sport." The Conversation, May 9. https://theconversation.com/ten-ethical-fl aws-in-the-caster-semenya-decision-on-intersex-in-sport-116448; Semenya, "Running in a Body That's My Own"; See also Jordan-Young and Karkazis, *Testosterone*.

57. T. A. Roberts, J. Smalley, and D. Ahrendt. 2021. "Effect of Gender Affirming Hormones on Athletic Performance in Transwomen and Transmen: Implications for Sporting Organisations and Legislators." *British Journal of Sports Medicine* 55: 577–83.

58. V. Seibert. 2014. "Michael Phelps: The Man Who Was Built to Be a Swimmer." *The Telegraph*, April 25. https://www.telegraph.co.uk/sport/olympics/swimming/10768083/Michael-Phelps-The-man-who-was-built-to-be-a-swimmer.html.

59. E. Pells. 2023. "Track Bans Transgender Athletes, Tightens Rules for Semenya." Associated Press, March 23. https://apnews.com/article/transgender-track-semenya-f3499b00b932948f96838adb3b010f11.

60. See for example D. Lopiano. 2022. "A Fair and Inclusive Solution for Transgender Women in Sports." *Forbes*, August 4. https://www.forbes.com/sites/donnalopiano/2022/08/04/a-fair-and-inclusive-solution-for-transgender-women-in-sports/?sh=7bdd75d012ef; and S. Erikainen, B. Vincent, and A. Hopkins. 2022. "Specific Detriment: Barriers and Opportunities for Non-Binary Inclusive Sports in Scotland." *Journal of Sport and Social Issues* 46 (1): 75–102. https://doi.org/10.1177/0193723520962937; and B. Hamilton, A. Brown, S. Montagner-Moraes, C. Comeras-Chueca, P. G. Bush, F. M. Guppy, and Y. P. Pitsiladis. 2024. "Strength, Power and Aerobic Capacity of Transgender Athletes: A Cross-Sectional Study." *British Journal of Sports Medicine* 58 (11): 586–97. https://doi.org/10.1136/bjsports-2023-108029.

61. "The Global Players' and Athletes' Association for Professional Sport—UNI World Athletes Explained." Law in Sport, August 25, 2015. https://www.lawinsport.com/topics/features/item/the-global-players-and-athletes-association-for-professional-sport-uni-world-athletes-explained#.

62. J. L. Herman, A. R. Flores, and K. K. O'Neil. 2022. "How Many Adult and Youth Identify as Transgender in the USA?" UCLA School of Law, Williams Institute. https://williamsinstitute.law.ucla.edu/publications/trans-adults-united-states/.

63. Jordan-Young and Karkazis, *Testosterone*; but see Hooven, *T: The Story of Testosterone*, for opposite claims.

64. K. Nakajima and C. H. Jin. 2022. "Bills Targeting Trans Youth Are Growing More Common—and Radically Reshaping Lives." NPR, November 28. https://www.npr.org/2022/11/28/1138396067/transgender-youth-bills-trans-sports.

65. The Trans Legislation Tracker. https://translegislation.com/.

66. G. R. Murchison, M. Agénor, S. L. Reisner, R. J. Watson. 2019. "School Restroom and Locker Room Restrictions and Sexual Assault Risk among Transgender Youth." *Pediatrics* 143 (6): e20182902. https://doi.org/10.1542/peds.2018-2902; L. Z. DuBois, J. A. Puckett, D. Jolly, S. Powers, T. Walker, D. A. Hope, R. Mocarski, et al. 2024. "Gender Minority Stress and Diurnal Cortisol Profiles among Transgender and Gender Diverse People in the United States." *Hormones and Behavior* 159: 105473. https://doi.org/10.1016/j.yhbeh.2023.105473.

INDEX

A NOTE ON THE TYPE

This book has been composed in Arno, an Old-style serif typeface in the classic Venetian tradition, designed by Robert Slimbach at Adobe.